DETECTION, CONTROL, and RENOVATION of CONTAMINATED GROUND WATER

Proceedings of a symposium sponsored by
the Committee on Water Pollution Management of
the Environmental Engineering Division
of the American Society of Civil Engineers
in conjunction with the ASCE Convention in Atlantic City, New Jersey

April 27-28, 1987

Co-sponsored by the United States Environmental
Protection Agency, Office of Ground Water Protection

Edited by Norbert Dee, William F. McTernan, and Edward Kaplan

Published by the
American Society of Civil Engineers
345 East 47th Street
New York, New York 10017-2398

ABSTRACT

The papers in this book provide technical guidance to engineers regarding the problems, solutions, and uncertainties associated with groundwater contamination in the United States. The papers are divided into the following areas: groundwater contamination, monitoring and detection, control of contamination, and renovating contaminated groundwater. Current knowledge as well as present difficulties are discussed. The role of the classically trained civil engineer in the evolving field of groundwater contamination is evaluated. Future directions for the field and an assessment of the additional skills needed to address these problems are also reviewed.

Library of Congress Cataloging-in-Publication Data

Detection, control, and renovation of contaminated ground water.

Includes indexes.
1. Water, Underground—Pollution—United States—Congresses. I. Dee, Norbert, II. McTernan, William F. III. Kaplan, Edward, 1943-   . IV. American Society of Civil Engineers. Environmental Engineering Division. V. United States. Environmental Protection Agency. Office of Ground-Water Protection.
TD223.D48   1987   363.7'394   87-1232
ISBN 0-87262-595-8

PREFACE

The papers contained in this Proceedings represent an overview of the status of ground water contamination in the United States. This Proceedings represents ASCE's initial formal effort in the area of ground water protection, and hopefully will lead to further understanding of the broad technical aspects concerning ground water contamination and protection. In that context, the papers address four topics:

- Ground Water Contamination
- Monitoring and Detection
- Control of Contamination
- Renovating Contaminated Ground Water

The papers document both the research that is taking place in ground water and the practical solutions to ground water problems using case studies as examples.

These papers have been accepted for publication by the Proceedings Editors, and are eligible for discussion in the Journal of Environmental Engineering Division and for ASCE Awards. We would like to thank the authors for their contribution to this Proceedings, that can serve an important reference for the status of ground water protection for both engineers and the general public.

Norbert Dee, PhD, ASCE
EPA Office of Ground Water Protection

William F. McTernan, PhD, ASCE
Oklahoma State University

Edward Kaplan, PhD, ASCE
Brookhaven National Laboratory

ACKNOWLEDGMENTS

The following individuals assisted in the preparation of this document as well as in the presentations made during the two-day symposium by serving as either technical reviewers and/or as session coordinators. Their efforts went far to help make this a successful symposium. Their work is acknowledged and appreciated:

Mr. Dennis D. Beckman
Woodward-Clyde Consultant
5055 Antioch Road
Overland, Park, Kansas 66203

Ms. Wendy L. Cohen
2320 Goldberry Lane
Davis, California 95616

Dr. Jeff Greenfield
Assistant Professor
Drinking Water Research Center
Florida International University
Miami, Florida 33199

Mr. Thomas C. Greengard
13682 W. Utah Circle
Lakewood, Colorado 80228

Mr. William Hoffstetter
Golder Associates, Inc.
4104 148th Avenue, N.E.
Redmond, Washington 98052

Dr. Nancy E. Kinner
Assistant Professor
University of New Hampshire
College of Engineering and
  Physical Sciences
Kingsbury Hall
Durham, New Hampshire 03824-3591

Mr. Darrel J. Kost, P.E.
State of New York
Department of Transportation
Veterans Memorial Highway
Hauppauge, New York 11788

Dr. G. V. Madabhushi
California Regional Water
  Quality Control Board
Santa Ana Region
6809 Indiana Avenue, Suite 2000
Riverside, California 92506

Dr. Jerry Overton
Department of Geology
Oklahoma State University
Stillwater, Oklahoma 74074

Mr. Michael A. Powers, P.E.
Goldberg, Zoino & Associates
320 Needham Street
Newton Upper Falls, Massachusetts 02164

Dr. Chet A. Rock
Department of Civil Engineering
University of Maine
455 Aubert Hall
Orono, Maine 04469-0105

CONTENTS

Session Program

*Manuscript not available at time of printing.

CHAPTER 1.0

SESSION PROGRAM

DETECTION, CONTROL AND RENOVATION OF CONTAMINATED GROUNDWATER

MONDAY MORNING
APRIL 27, 1987

Session No. 2-S2.1 ....9:00 a.m.

GROUNDWATER CONTAMINATION

Environmental Engineering Division

<u>Presiding</u>:   NORBERT DEE, US EPA, Washington, DC

9:00   Introductory Remarks

9:10   Keynote Address

9:50   Extent of Groundwater Contamination in the U.S.: An Overview
MARIAN MLAY AND NORBERT DEE
Office of Groundwater Protection, US EPA, Washington, DC

10:10   Engineering Aspects of Point vs Non-Point Sources of
Groundwater Pollution
J. DAVID DEAN AND DAVID R. GABOURY
Woodward-Clyde Consultants, Walnut Creek, CA

10:30   Survey of Groundwater Contamination in Massachusetts
STEVEN P. ROY
Groundwater Protection Dept. DEQE, Division of Water Supply,
Boston, MA

10:50   Synthetic Organic Contaminants and Pesticides in Groundwater:
An Emerging Health Problem
MAHFOUZ H. ZAKI
Suffolk County Department of Health Services, Hauppauge, NY

11:10   A Rational Framework for National Groundwater Management
JOHN GASTON
Department of Environmental Protection, Trenton, NJ

Session No. 9-S2.2 ....2:00 p.m.

MONITORING AND DETECTION

Environmental Engineering Division

Presiding: G. LEE CHRISTENSEN, Villanova University, Villanova, PA

2:00   Overview of Groundwater Monitoring Techniques
         RICHARD SCHOWENGERDT
         International Technology Inc, Milwaukee, WI

2:20   Technical Issues of Groundwater Data
         OLIN BRAIDS, GISELLA M. SPREIZER and GREGORY SHKUDA
         Geraghty & Miller Inc, Plainview, NY

2:40   A Call for New Directions in Drilling and Sampling Groundwater
         Monitoring Wells
         JOSEPH F. KEELY
         Oregon Graduate Center, Beaverton, OR

         KWASI BOATENG
         Roy F. Weston, Inc., West Chester, PA

3:00   Managing Groundwater Data
         EDWARD KAPLAN and ANNE F. MEINHOLD
         Brookhaven National Laboratory, Upton, NY

3:20   Techniques for Delineating Subsurface Organic Contamination:
         A Case Study
         ANN M. PITCHFORD and ALDO T. MAZZELLA
         EMSL, US EPA, Las Vegas, NV

         EDWARD HEYSE
         Air Force Engineering Services Center, FL

3:40   PANEL DISCUSSION: Producing Useful Data

         Participants: Olin Braids (Geraghty & Miller, Inc.)
                        David Dean (Woodward-Clyde)
                        Kwasi Boateng (Roy F. Weston, Inc.)
                        Ann Pitchford (EPA/EMSL)
                        Richard Schowengerdt (ITI)

EXTENT OF GROUND WATER CONTAMINATION IN THE U.S :AN OVERVIEW

Marian Mlay[1], J.D. and

Norbert Dee[2], Ph.D., ASCE

Abstract

Our society depends upon an available supply of clean water. Ground water provides about half the drinking water in the U.S. and is extensively used in industry and agriculture. The natural quality and quantity of ground water varies through the nation. Instances of ground-water contamination have been found in most sections of the country.

The contamination of ground water may result from all aspects of human activities: agriculture, industry, transportation, domestic wastes and resource exploitation. The contamination due to man has occurred for centureies, but urbanization, industrialization and increase population density have greatly aggravated the problem in some areas. The contaminants found in ground-water vary from simple inorganic ions to complex synthetic organic chemicals.

Public perception of ground-water protection as a significant environmental issue is relatively recent, evoling over the past decade. Previously, the focus was on pollution we could see and smell. Moreover, both experts and the public alike believed that ground water was self-cleansing. In the late 1970s and early 1980s, uncontrolled waste sites such as Love Canal and Valley of the Drums, together with pesticide incidents such a EDB and growing reports of chemical-tainted drinking water, slowly focused attention on ground water.

Ground-water protection is by far the most complex of virtually all environmental management issues. In magnitude it involves of tens of thousands of individual sources and an enormous array of domestic, commercial and industrial practices. The potentially regulated community encompasses not only a few large industries and businesses but also small businesses, homeowners and agriculture.

[1] Director, Office of Ground Water Protection, U.S. EPA, Washington, D.C.

[2] Senior Scientist, Office of Ground Water Protection, U.S. EPA, Washington, D.C.

The technologies for determining actual ground-water quality as well as for preventing and cleaning up contamination are becoming more sophisticated but still are inadequate for the enormity of the job. Moreover, environmental management is only gradually becoming ground-water oriented. Major environmental statutes, such as the Safe Drinking Water Act, Resource Conservation and Recovery Act and Federal Insecticide, Fungicide and Rodenticide Act, now are being used for ground-water protection, although initially that was not their principal purpose.

In recent years, government at all levels has made substantial progress in addressing ground-water protection, but much still remains to be done. States continue to have the primary responsibility for managing ground-water quality, and many have adopted ground-water quality standards or classification systems and have stepped up monitoring efforts. In addition, the wellhead protection program established by the Safe Drinking Water Amendments of 1986 will also be used by many States to protect their ground water.

This paper provides an overview of why ground-water protection has become such an important issue in recent years and describes future trends in accomplishing ground-water protection. It discusses the importance of the ground-water resource, the nature and extent of contamination, the health and environmental impacts of contamination, and the continuing Federal, state, and local efforts to protect the quality of ground water.

IMPORTANCE OF GROUND WATER

Ground water is continuing to grow in importance, not only as a source of drinking water, but also for agricultural and industrial needs. Ground water makes up about one-fourth of all the fresh water used in the United States. Between 1950 and 1980, total ground-water withdrawals increased from 34 to 89 billion gallons per day (BGD), an increase of 162 percent. The 1980 figure represents 24 percent of all the fresh water used (372 BGD) that year. In part, this increase has been the result of changes in irrigation and population migration during the 1970s to rural and suburban areas, where ground water is more easily accessible than surface water. The 1985 ground-water withdrawals were projected to reach 100 BGD.

About 117 million people in the United States rely on ground water for their domestic needs. Of the 100 largest cities, 34 derive their water either completely or partly from ground water. And in the seven most populated states -- New York, California, Florida, Illinois, Ohio, Michigan, Texas, and New Jersey -- more than 52 million people receive their drinking water at least partly from ground water. Of the 622 public water supply systems in New Jersey, 558 obtain most of their supplies from ground water. In the less populated, rural areas of the country, 95 percent of the residents depend entirely on this resource for domestic uses.

The agricultural states in the Midwest and the West depend heavily on ground water for irrigation. Arkansas, Nebraska, Colorado, and Kansas use over 90 percent of their ground water for agricultural activities.

Although small when compared with the quantities of ground water used for agriculture, some states' withdrawals for industrial uses constitute a large portion of their total withdrawals. Because a significant number of industries are located in the eastern half of the country, many states there use over 30 percent of their ground water for industrial purposes. Kentucky uses 58 percent of its ground water for industry.

In summary, ground water presently supplies half of the nation's drinking water; it accounts for 35 percent of the municipal drinking water supply and about 95 percent of that in rural areas.

NATURE AND EXTENT OF GROUND-WATER CONTAMINATION

Although we are constantly expanding our understanding of the extent of the ground-water problem, we may never have a complete picture of the nature and extent of contamination. Our current data are fairly limited, and ground-water monitoring is very expensive and technically difficult.

In addition, the lack of a clear understanding of movement of contaminants in ground water means that even a sophisticated monitoring system could fail to detect existing contamination. Water from one well can be uncontaminated, while that in a well only ten feet away may be contaminated. Finally, ground water is much less accessible for testing than surface water, and its movement and mixing properties are very slow compared to those of surface water. Testing of ground water involves identification of the proper location and depth, drilling and properly screening a well, and then conducting sophisticated sampling procedures.

Lacking scientific ground-water monitoring data, anecdotal data can be used to indicate problems. The best indicator is the drinking water which is drawn from ground water. Twenty percent of all drinking water systems and 30 percent of those in municipal areas show at least trace levels of VOC or man made contamination.

TYPES OF CONTAMINANTS

Currently, the contaminants that are of major concern are the man-made "toxic" chemicals. There are, however, many other types of contaminants that are also of concern, such as natural, microbiological and other conventional pollutants. In 1984, the Office of Technology Assessment (OTA) listed 175 organic chemicals, over 50 inorganic chemicals and various biological organisms and radionuclides that have been found in ground water throughout the nation; this is still not likely to be a complete inventory. (OTA, 1984).

Biological contaminants include bateria viruses, parasites, and other agents of illness. Although information available about the pathogens which are actually present in ground water are limited, the OTA found reports of typhoid, tuberculosis, cholera and hepatitis. During the past 10 years, ground water serving over 100 million people, or 10 percent of public ground water supplies, reportedly exceeded EPA's drinking water standard for microbiological contaminants. Small systems had the most prevalent violations. EPA's own rural water survey also detected serious microbiolgical problems, with 30 percent of the 2,700 groundwater systems tested exceeding the standard for microbiological contaminants in drinking water. (EPA, 1985)

Inoganic contaminants include salts, metals and other substances that do not contain any carbon. OTA has identified 37 inorganics, including 23 metals, known to be present in ground water. Over the past decade, approximately 1,500 to 3,000 ground water supplies have exceeded the primary or secondary drinking water standards which EPA has established for 13 inorganic substances. Much of the inorganic contamination is, however, from natural causes.

Other conventional types of contaminants include chlorides, sulfates, nitrates and metals, some of which are natural, some man made. Nitrates, for example, appear to be the most common contaminant.

Organic substances, particularly synthetic organic compounds developed over the past 50 years, are of growing concern. OTA reported that 175 different organic chemicals have been found in ground water, and that the concentrations

of these chemicals in ground water is substantially higher
than in surface water. EPA ground-water surveys conducted
over the past decade confirm the ubiquitous nature of
organics in ground water. The pervasiveness of these substances is largely due to the fact that they are commonly
used in daily life. Solvents, pesticides, paints, dyes,
varnishes, and ink are just a few examples of products
containing organic chemicals.

SOURCES

     The sources of contamination are many and varied. We
are just beginning to understand that the highly publicized
waste sites are not the main source; non-waste practices
may, in fact, account for up to two-thirds of the ground
water contamination.

     There are many ways of grouping these numerous sources.
OTA in its report divided 33 sources into six categories.
Two additional sources of contamination which should be
added to the 33 sources are nuclear facilities and abandoned waste sites. (Table 1)

TABLE 1 - SOURCES OF GROUNDWATER CONTAMINATION

Category I - Sources designed to discharge substances
Subsurface percolation (e.g., septic tanks and cesspools)
Injection wells
Land applicaiton

Category II - Sources designed to store, treat, and/or
dispose of substances; discharge through unplanned release
Landfills
Open dumps, including illegal dumping (waste)
Residential (or local) disposal (waste)
Surface impoundments
Waste tailings
Waste piles
Materials stockpiles (non-waste)
Graveyards
Animal burial
Aboveground storage tanks
Underground storatge tanks
Containers
Open burning and detonation sites
Radioactive disposal sites

Category III - Sources designed to retain substances during
transport or transmission
Pipelines
Materials transport and transfer operations

Category IV - Sources discharging substances ad consequence
of other planned activities
Irrigation practices (e.g., return flow)
Pesticide applications
Fertilizer applications
Animal feeding operations
De-icing salts applications
Urban runoff
Percolation of atmospheric pollutants
Mining and mine drainage

Category V - Sources providing conduit or inducing discharge
through altered flow patterns
Production wells
Other wells (non waste)
construction excavation

Category VI - Naturally occurring sources whose discharge
is created and/or exacerbated by human activity
Groundwater-surface water interactions
Natural leaching
Salt-water intrusion/brackish water upconing (or
intrusion of other poor quality natural water)

SOURCE: Office of Technology Assessment, 1984

    Examples of improper waste disposal contamination
sources include an estimated 20,000 - 22,000 hazardous
waste sites which are candidates for the Superfund National
Priorities List; there are approximately 1,000 sites currently on the list. An estimated 22 million septic systems
are sited to leach, and over 90 percent of the 181,000 surface impoundments pose a ground-water contamination threat.
Finally, there are approximately 93,000 landfills, mostly at
industrial sites, which are sources of contamination.

    The non-waste category includes both point and non-
point sources. The major non-waste contamination sources
are underground storage tanks and pesticides. There are
four to five million underground storage tanks for petroleum and non-petroleum substances, of which an estimated 30
percent leak. Ground-water contamination for 17 pesticides
has been discovered in 22 states, both from normal agricultural use and more from spills and improper disposal.
Other non-waste practices include the production and use of
herbicides and fertilizers; highway deicing, chemigation,
road coatings and drawn downs/salt water intrusion. (Table
2)

TABLE 2

Number of States Identifying These Major
Sources of Ground Water Contamination

Septic Tanks	36
Municipal landfills	31
On Site Industrial Landfills	30
Other Landfills	20
Surface Impoundments	36
Oil and Gas Burn Pits	21
Underground Storage Tanks	41
Injest on Wells	18
Abandoned Hazardous Waste Sites	25
Regulated Hazardous Waste Sites	15
Salt Water Intrusion	19
Land Application	7
Agricultural	33
Highway De-icing	11
Other	35

Source OGWP 1985

The relative seriousness of each source depends on
many factors, and even experts in the field disagree as to
which sources pose the greatest threat. The factors which
impact the degree of severity of the contamination include:

-   What is released; the degree of toxicity of the
    substance and whether or not it is biologically or
    chemically modified;

-   The amount released; in general, small amounts cause
    less of a problem, but small amounts from many
    sources, such as septic tanks, can pose a greater
    threat;

-   When it is released; for example, a one-time spill
    creates problems which are different from those
    created by a continuous release; and

-   Where the contaminant is released, both hydrogeologi-
    cally and in relation to human water use patterns.

The one thing that the experts can agree on is that
the potential for ground-water contamination is widespread
and that non-point, non-waste and non-hazardous sources
join the infamous hazardous waste sites as significant
sources of contamination of our Nation's ground water.
(OTA 1984)

EFFECTS OF GROUND-WATER CONTAMINATION

The principal concern over ground-water contamination is its potential impact on human health, but concern with environmental effects is growing. One-half of the population of the United States, or 117 million people, use ground-water as a source of drinking water. Ground-water supplies 48,000 of the 60,000 existing public water systems, as well as 12 million individual wells. The ground water which supplies public water systems is generally not treated; there is a disinfection process, but this prevents only microbiological contamination.

Health Effects

Health effects from biological contaminants have long been well-documented, but knowledge of the effects of organic, inorganic and radioactive substances in drinking water still has some gaps. The populations served by untreated ground-water supplies have 3.7 times as many cases of illness from biological contaminants as those served by treated ground water supplies. Gastrointestinal and viral outbreaks are the principal manifested health effects. These outbreaks due to ground-water contamination have been growing over the last 15 years from eight to 17 per year, and illness related to ground-water contamination account for 28 percent of all reported water-borne diseases.

Many organic chemicals are known to be both toxic and carcinogenic. Toxic effects may include impairment of the liver, kidney and central nervous system. At the present time, however, no data is available on the toxicity or carcinogenic potency of numerous organics found in ground water.

Inorganic contaminants such as metals, lead and mercury can cause both acute and toxic effects. Nitrates can cause "blue baby"; and evidence is growing that it is also a causative factor in cancer, neurological disorders and birth defects. Radon, a radionuclide, causes an estimated 500 cancer deaths per year by ingestion and up to 10,000 deaths annually by inhalation.

Environmental Effects

Although the principal concern over ground-water contamination is related to its adverse impact on human health, concerns with its environmental and ecological effects is also growing. Documentation of these impacts remains limited, however. Potential impacts of ground-water contamination include adverse effects on surface waters and damage to fish, vegetation and wildlife. For example, 15% of endangered species are ground-water related. (OGWP 1986)

STATE AND FEDERAL ROLES IN GROUND-WATER PROTECTION

The enormity of the task of protecting the nation's ground-water resources requires a combined State, local and Federal effort. Primary responsibility rests with the States and localities and is historically linked to water rights as well as land use and development issues and policies.

The Federal role has been, and continues to be, related to its statutory authorities to control contaminants, such as pesticides and high level radiation wastes; sources of contaminants, such as hazardous waste sites; certain practices, such as underground injections; and certain uses, such as drinking water quality. In a survey of EPA's program offices conducted by Office of Ground Water Protection (OGWP), EPA program offices reported some activity for 25 of 33 sources identified by OTA. Most, however, of the activities are related to waste management and regulations (OGWP, 1987)

At first, state efforts were almost entirely reactive and were related almost exclusively to incidents of well contamination. In recent years, however, states and localities have been significantly increasing the attention they give to ground-water protection. The National Governors' Association policy statement issued in 1986 is one example of this strenghtened commitment at the state level. State and local efforts aimed at accepting and implementing the delegation of certain Federal programs, such as RCRA, UIC and Superfund, are further evidence of the greater importance attached to ground-water protection efforts.

The current trend in ground-water protection is now towards the development of state ground-water strategies, well head protection legislation and institutions. States and localities have also taken a more active role in land use control as a means of protecting the ground-water. Two states, New Jersey and Texas, have reclaimed control over two areas where there were threats to ground-water quality. In New York State, Suffolk County established zoning regulations which made 45 percent of the county off-limits to industry and waste handlers; and Dade County, Florida established well field protection zones eight to ten miles around the county well fields. States are also active in monitoring and collecting data about ground-water quality. At the present time, thirty-eight states conduct some type of ambient monitoring.

FUTURE TRENDS

Effective long-term protection of our ground-water resources requires us to confront highly challenging technical issues and public policy questions. We must obtain more and better information, conduct more research, develop new techologies and make decisions on the nature and extent of protection for which we are willing to pay.

References Cited

National Statistical Assessment of Rural Water Conditions (U.S. Environmental Protection Agency, Office of Water, Office of Drinking Water, June 1984)

Protecting the Nations Ground Water from Contamination (Washington, DC; US. Congress, Office of Technology Assessment, OTA-0-233, October 1984)

Overview of State Ground Water Program Summaries (U.S. Environmental Protection Agency, Office of Water Office of Ground Water Protection, March 1985)

Guidelines for Ground-Water Classification under the EPA Ground-Water Protection Strategy (U.S. Environmental Protection Agency, Office of Water, Office of Ground Water Protection, 1986)

Agency Activities Related to Sources of Ground-Water Contamination (U.S. Environmental Protection Agency, Office of Water, Office of Ground Water Protection, 1987)

Engineering Aspects of Point Versus
Nonpoint Sources of Groundwater Pollution

J. David Dean* and David R. Gaboury**

Introduction

The importance of groundwater as a resource for maintaining life and human productivity cannot be overstated. According to a report of the U.S. Environmental Protection Agency in 1977, approximately one half of the population of the United States depends upon groundwater for its potable water supply. Supplies of groundwater are not inexhaustible, however, and there are alarming numbers of references documenting the contamination of the existing resource (Todd and McNulty, 1976; Freeze and Cherry, 1979).

Pollution of groundwater may occur due to a number of sources. These may be categorized into point and nonpoint sources of contamination. The purpose of this paper is to explore the issues associated with the engineering aspects of these two source types. The paper will begin by giving a brief historical perspective on why these sources of contamination have come into existence. Next, point versus nonpoint sources will be defined and the motivation for making a distinction between them stated. The remainder of the paper will discuss the engineering issues associated with prediction and control of pollutant loadings from these sources to groundwater.

A Brief Historical Perspective on Causes of Groundwater Pollution

The soil has long been thought of as an ultimate disposal resource. Landfills, cesspools, and "garbage dumps" have been utilized for as long as mankind has been dwelling in permanently settled areas. Disposal into water bodies or onto the land has always been an expedient means of dealing with human waste.

Sanitary waste disposal into rivers and lakes received the attention of sanitarians first. Nuisance or, at times, linkage of contaminated water to outbreak of epidemic disease forced the need for alternate disposal measures. A natural alternative to disposal of sanitary wastes into surface waters was surface disposal onto or burial in the soil. This practice has continued unabated and uncontrolled in the United States until recent times.

An as example of the historical engineering perspective on the land disposal of liquid and solid wastes and its impacts on groundwater, we offer the following observations from engineering texts on the treatment

*Woodward-Clyde Consultants, 100 Pringle Avenue, Suite 300,
Walnut Creek, CA 94596.
**ASCE Member

and disposal of wastewater. Our selection of text is arbitrary and is only intended to illustrate the state of the practice over the last twenty or so years.

As late as the 1960s, there was little or no consciousness of the finite capability of the soil to purify percolating water. Fair and Geyer (1963) stated that "self purification of groundwaters departs significantly from that of surface waters. The variety of living organisms that seize upon the pollutional substances for food is greatly restricted in the confinement and darkness of the pore space of the soil. But this reduction in biological forces is more than counter-balanced by the introduction of the physical force of filtration. In general, the rate of purification is stepped up greatly, and the time and distance of pollutional travel shrink to smaller values." This viewpoint changed over the next decade. Metcalf and Eddy (1972) mentioned the disposal of sewage treatment effluent on land and for groundwater recharge, without mention of quality considerations. In regard to sludge disposal in landfills, they offered the single caveat "Drainage from the site that would cause pollution of groundwater supplies or surface streams should be guarded against." Many of the present sources of groundwater contamination were engineered specifically to promote percolation of the waste through the soils (e.g., septic tanks, percolation ponds). Gehm (1976) discussed disposal of municipal sludge in lagoons. In this publication it is stated that subsoil permeability is involved in lagoon design and that "Permeable bottoms are desired and should be at least 18 in. above the maximum water table". WPCF and ASCE (1977) recognized depth to groundwater as a design issue and provided the following commentary on land application of wastewater, "Because of the minimal amount of experience pertaining to groundwater degradation via land application systems, the design engineer should be aware that the irrigation methodology may cause groundwater degradation". Thus, we see that as little as a decade ago, there was no standard in the practice for evaluating the potential degradation of groundwater by sanitary waste disposal.

The potential for groundwater pollution from industrial and hazardous waste disposal similarly received little attention until the 1970s. Besselievre (1969), to the knowledge of the authors, does not mention the potential for groundwater pollution in his text on the treatment of industrial wastes. Only in one instance, the case of oxidation ponds, is protection of groundwater hinted at. He states "where soils are permeable it is deemed good practice to line lagoons and oxidation (ponds) with plastic membranes or asphalt or other materials." Hagerty, et al. (1973), describe a ranking system for selection of sites for hazardous waste landfills. In this system, bottom leakage potential, filtering capacity and adsorption capacity of subsurface soils, the buffering and microbiological degradation capabil-ities of groundwater systems, as well as travel distance and groundwater velocites are considered. Most recently, Overcash and Pal (1979) present a design text for the treatment of industrial waste that is based upon the non-degradation of the terrestrial receiving system which acknowledges the potential for and protects against pollution of soil and groundwater. Thus, we see that it is only in the past 10-15 years that the problem of groundwater pollution from industrial disposal has been recognized and treated as a design consideration.

Another cause of groundwater contamination has been the response to other regulatory pressures. For instance, the "zero discharge" language of the Clean Water Act of 1976 made land disposal of sanitary, municipal and industrial wastes even more attractive. In the implementation of design to attain this concept, however, there was little consciousness of the potential contamination of groundwater by these activities, and few measures were taken to insure against it. Another example of regulatory pressure is in the case of underground storage of fuels and flammable solvents. Such facilities were originally forced underground because of fire codes.

In defense of the pollution control engineer, it can be said that the problem of groundwater contamination has come only recently to our attention due to our recently enhanced capability to detect it. Analytical capabilities have increased substantially over the past two decades. In the case of cadmium, for example, the use of the graphite furnace AA technique introduced around 1974 decreased the detection limit to about 3 parts per trillion over the conventional flame AA technique (detection limit approximately 1 ppb) which was the best available detection prior to that time. In the case of organics, the flame ionization detector used for gas chromatography in the 1960s had a detection limit of about 20 ppb. Electron capture, introduced in the mid-seventies brought that limit to about 0.01 ppb. Since that time the detection limit has fallen another factor of 10 with the introduction of yet more sensitive methods.

Because computed health effects levels are, at yet lower concentrations, there is a continued motivation to improve detection limits. As these analytical methods continue to improve, so will the discovery of contamination in groundwater previously considered clean.

We are continuing to develop our understanding of the soil and groundwater as a self-regenerative waste disposal system. As will be discussed in more detail in this paper, that body of knowledge is far from complete. It is important and appropriate that our lack of understanding about these systems be matched by an appropriate degree of conservatism in design and regulation writing.

Definition of Point Versus Nonpoint Sources

In order to discuss the issues concerning point and nonpoint sources of groundwater contamination we need a working definition which categorizes them. Historically, the concept of nonpoint pollution source arose from the consideration of surface waters. A point source was classified as a concentrated discharge from a pipe directly into the receiving water. A nonpoint source was, in essence, considered to be anything and everything which did not fit the definition of a point sources. As the technology evolved for dealing with nonpoint sources, however, they were endowed with characteristics of their own. Several authors have attempted to define nonpoint sources (Overcash and Davidson, 1980; Novotny and Chesters, 1981); however, these definitions were tailored to surface water, rather than groundwater sources.

The chief components of a definition of a nonpoint source would seem to entail the following:

o   The pollutant releases from the source are discontinuous in time and are driven by natural hydrological processes,

o   The source is spatially diffuse and exact entry points into the receiving water cannot be easily defined.

The implications of this definition are interesting to note. The fact that the loadings occur intermittently in time and space implies that nonpoint source loadings to groundwater are difficult or impossible to measure and that they are largely uncontrollable once the releases begins. This implies that the principal motivation for categorizing sources is a regulatory one. One can easily see that the methods for regulating sources which can be measured and controlled would be quite different from sources which can not. For engineering purposes, there is no strong motivation for categorizing these sources. In fact, many of the same technical problems may be involved for both source types. This will become evident as we discuss the issues.

Pollution Sources and Characterization

A list of potential sources of groundwater contamination is shown in Table 1. The list is not exhaustive. Each source type is associated with a list of major waste constituents and a categorization.

Categorization of these sources by the definition given above is not straightforward. In fact, one of the few source types that truly qualifies as a point source of groundwater contamination is an injection well. Such a well clearly meets our criteria of a point source because the loading to groundwater is both directly measurable and controllable. A source type which is clearly nonpoint is agricultural leachate. In this case, the loadings are both driven by the hydrologic regime and are diffuse. In some instances, however, such as in the case of solid waste disposal sites, the distinction is not so clear. While such sites, in general, may have characteristics of a nonpoint source (i.e., loadings may be discontinuous and hydrologically driven), for any given site the point of entry is well defined and (at least for new sites) engineering controls can be implemented to control leachate. If one considers the case of a region which may have a number of abandoned, uncontrolled landfills, then the source might be considered nonpoint. This may also be the case for waste lagoons, sewage and wastewater disposal sites and some waste piles. In such cases, the categorization of these source types may depend upon whether they are controlled or uncontrolled.

The remainder of the paper will discuss the issues associated with the engineering aspects of both of these source types. The engineering problems break down into three basic categories: prediction of pollutant loading to groundwater, design of engineering controls to prevent or detect releases and remediation of releases. Because of the limitations in space, remediation issues will not be covered. Many of the issues involved in remediation are independent of the source. Even

POTENTIAL SOURCES OF GROUNDWATER CONTAMINATION

Source	Major Constituents[1]	Type[2]
Septic Tanks	B, N, I	NP
Sewage and Wastewater Disposal	B, N, P, S, O, I	P
Agricultural Leachate	B, N, P	NP
Solid Waste Disposal	N, P, O, M	P
Chemical Spills	N, P, S, PP, O	NP
Deep Well Injection	PP, O, I, M	P
Deicing Salts	I	NP
Mine Leachate	O	NP
Waste Lagoons	N, P, S, PP, O, M, I	P
Waste Piles	O, M, I	P
Artificial Recharge	N, P	P
Underground Tanks	S, PP	P
Abandoned Dry Wells	All	NP

[1] B - Bacteria, Viruses
     N - Nutrients
     P - Pesticides
     S - Solvents
     PP - Petroleum Products
     O - Other Toxic/Carcinogenic Organics
     M - Metals
     I - Miscellaneous Inorganics

[2] NP - Nonpoint
     P - Point

though these are a significant number of unresolved problems with
existing models, as will be discussed below they represent an important
aid for designing disposal systems for wastes in a complex environment.

Prediction of Pollutant Loads

By the definitions which we have stated, direct engineering controls
of nonpoint sources of groundwater pollution are virtually nonexist-
ent. Unlike controls for nonpoint sources of surface water pollution
(e.g., sedimentation basins) a posteriori controls of nonpoint groundwater sources cannot be feasibly implemented. Controls which can be
implemented to control nonpoint sources include regulatory and incentive
measures such as restriction of use, permits and taxation. However,
regulatory controls such as use restriction should be based on engineering calculations. Otherwise, chemical manufacturers, distributors and
users may unjustifiably be penalized, or the restriction may not be
adequate to protect the resource. Likewise, fail-safe controls for
point sources (e.g., limitation of disposed chemical mass or concentration) should be based upon the ability of the soil and hydrogeologic
system to dilute or detoxify uncontrolled releases to appropriate
levels. An appropriate framework for designing such restrictions would
be a "waste load allocation" type approach, either for nonpoint sources
or widely distributed point sources, especially if the chemicals
involved have long half lives.

The design of restrictions should be based on predictive chemical
fate and transport modeling. Moreover, they should take into account
the risk to human health or loss of beneficial use of the resource based
upon the model results. For nonpoint sources, these models must be
capable of representing the discontinuous nature of the loading process
in time and space. Therefore, they should be capable of simulating
unsteady soil water hydraulics and contaminant transport and spatial
variability in these processes. In addition to providing the loadings
to the aquifer, they must be capable of simulating subsequent drinking
water well contaminant concentrations resulting from these loadings.

Currently, there are few, if any, models which are capable of simulating both the impact of management practices on loadings and the
resulting groundwater concentrations. Other problems with existing
models include their use of simplifying assumptions to facilitate
solution of the flow and transport equations, their sometimes
proprietary nature, lack of documentation, lack of user guidance in
parameter estimation, and lack of field validation.

Leaching of dilute concentrations of organics through the unsaturated zone has been well studied. Although nonlinearity and hysteresis
have been documented in both adsorption and desorption to soil materials, sorption is conveniently described by the use of the organic carbon
partition coefficient and the fraction of soil organic carbon in most
models. First-order approximations are normally used to describe degradation in soils. Current research issues include the quantification of
degradation or detoxification rates in the subsurface environment,
effects of dissolved organic carbon on facilitated transport and the
effects of macropores on water and solute movement. The transport of

volatile organics in the unsaturated zone deserves far more attention as the effects of vapor flow are not simulated routinely in currently available models.

Significant uncertainties exist in the prediction of ion transport in the unsaturated zone. Selectivity coefficients required to predict cation exchange are difficult to measure and are very dependent upon environmental conditions (e.g., Eh/pH, temperature, the nature of the exchange sites) and on the chemical composition of the exchange complex and the soil solution. Similar uncertainties exist in computing anion exchange and subsequent transport.

Although the dynamics of nutrients (N and P) in soils has been extensively studied, research is still ongoing to quantify the nitrogen balance. Significant uncertainties exist in the estimation of nitrogen transformation kinetics in various soils, denitrification and gaseous losses of elemental nitrogen and ammonia from feedlots and fertilized fields.

For some time, it was thought that the soil provided an effective filter for particulates such as bacteria and viruses. Recent work on in-situ biodegradation of organic contaminants in soil and groundwater point out that subsurface water can be abundant in microorganisms. The fact that beneficial organisms are in groundwater attests to the fact that the resource is susceptible to pollution by undesireable organisms as well. Further research in subsurface biodegradation will lead to a better understanding of survivability and transport of microorganisms in soil and groundwater.

Most nonpoint sources, as they have been defined here (with the exception of concentrated chemical spills), tend to involve dilute concentrations of chemical substances in water. In such cases, many modeling assumptions of convenience can be employed. In the case of chemical spills and some point sources (e.g., waste lagoons or underground tanks), however, these simplifying assumptions may not be applicable and prediction becomes both more difficult and uncertain.

Substantial work needs to be done where two phase (i.e., water and miscible and/or immiscible organic phases) are involved. While some models are available which describe the movement of several phases (e.g., compositional models) in the vadose zone, they are not used in general engineering practice. This class of problems can apply both to surface spills and to leaks of pure solvents or petroleum products from underground tanks. Large uncertainties exist in the biodegradation of pure phase chemicals in the unsaturated zone and the effects of high concentrations on the transport medium itself (e.g., the conductivity of clay materials exposed to high concentrations of solvents).

Underground injection wells pose some unique problems. The effects of high temperatures and pressures on chemical reactions among the constituents of the injected fluid and the ambient aquifer fluids are not well understood. The potential migration of these contaminants into overlying aquifers through fractures, leaky and/or discontinuous aquitards also poses a formidable engineering challenge.

Among the key issues related to such predictive modeling of loadings from both point and nonpoint sources is the uncertainty in contaminant and media properties. Uncertainties in basic properties of the contaminant which affect its movement to groundwater include its tendency to adsorb to soil materials, its degradation rate and/or the rate at which it is transformed to another substance, and chemical speciation tendencies. Likewise, uncertainties exist in pertinent properties of the media including hydraulic conductivity, pH, organic carbon fraction or cation exchange capacity, minerological makeup and microbiological status.

In the case of single chemicals, such as a pesticide applied over a large area, the principal source of uncertainty is in the description of the properties of the transport media. Current technical problems include not only describing the media properties, but also in making use of this information in modeling applications. Recently, a body of literature dealing with the spatial variability in transport-associated media properties is beginning to emerge. Reviews of this literature are available (Jury, 1985; Shields, et al., 1986).

Very few models are available which can account for the uncertainty in their parameters. Most of the truly stochastic models reside in the academic community and are not in general use in engineering practice. An alternative approach to the problem has made use of deterministic models and a Monte Carlo simulation strategy. Recent examples are Carsel, et al. (1985) and Dean, et al. (1986), using the Pesticide Root Zone Model (PRZM).

In the case of high concentrations of multiple chemicals, especially if the release occurs over a relatively small area, a larger source of uncertainty may reside in the characterization of the source material properties themselves.

Engineering Controls

In terms of controlling migration of pollutants to groundwater, there are two aspects to consider: prevention and detection.

Aquifer protection can be brought about through the implementation of physical engineering designs that will eliminate or minimize the migration of pollutants to groundwater or through judicious selection of disposal or treatment sites. In the case of solid waste disposal sites, the U.S. EPA is currently undertaking an integrated program of aquifer protection. The minimum technology (double liner design) standards are seen by the Agency as eliminating the threat of groundwater contamination during the operating life and post-closure period (30 years) for these units. Simultaneously, the Agency has undertaken to develop models which can be used to specify treatment levels (i.e., limitations on mass and/or concentrations) for disposed constituents, thereby seeking protection of the aquifer after the post-closure period or prior to that period in case of design failure (Federal Register, 1986). This procedure makes use of pertinent information about the landfill itself, the chemicals disposed, and the properties of the underlying aquifer system. The model is cast into a risk assessment framework and makes use of the uncertainty of these parameters on a site-to-site basis. Thus, given a fixed treatment level, the model could as easily be used to select sites in which the natural system affords some minimum level of protection. The model does not consider more than one landfill in an aquifer system, however, at this time. This model is currently being modified so that within site uncertainty in properties can also be accounted for. This extension will allow an evaluation of the risk involved in selecting a specific site for solid waste disposal. For most point sources, the risk at a particular site, not the risk associated with multiple sites in an aquifer system, is of greater importance.

There are several issues associated with the development of such protective measures. First, there is a certain probability that components of the landfill design (i.e., geomembrane, leachate collection system, clay liner) will fail. Some work on evaluating the likelihood of such failures has been done. This becomes less of an issue if siting of landfills and setting of minimum treatment standards for disposed contaminants to provide a minimum protection level is accomplished and enforced. However, in a waste load allocation framework, failure probability would still be an important issue. The issues associated with site selection or selection of treatment levels are much the same as for predictive modeling of loads from uncontrolled sources (e.g., the uncertainty in properties of the transport medium). Additional issues concern the characterization of the chemical properties of the source. Research has shown that the leachate from landfills contains a plethora of organic constituents which may react with or facilitate the transport of toxicants to groundwater. Some models to account for the interaction of components in simple, two or three mixtures, have been proposed.

The second aspect of controlling contaminant loadings to groundwater is through detection. A good source type as a case study to discuss the issues is provided by underground tanks which store solvents or petroleum products. In California, current minimum technology calls for the replacement of single walled with double walled containers. The space between the two walls provides (as in the leachate collection system of a landfill) an opportunity to detect leaks by the simple detection of the presence liquid. The vadose zone provides another opportunity for leak detection for volatile chemicals using soil gas monitoring. Finally, the groundwater beneath the tank can be monitored for the presence of contaminants.

Both of the first two detection methods involve the use electronic devices. Both involve continuous monitoring to detect a low probability event. Reliability of such systems over the relatively long life of a storage tank is an issue for which there is little or no confirming. While monitoring in the interwall space may detect tank leaks, it does not detect leakage in piping or fittings, which may be a greater source of leaks than tank failure itself. With vadose zone monitoring, there is the additional possibility of false positive detections induced by surface spills.

Groundwater monitoring may be the most reliable means of detection even though detection occurs after the fact. However, early detection can lead to minimization of an otherwise large-scale problem. As with vadose zone monitoring, surface spills may result in false positives. Concentrations found in groundwater may not be indicative of the magnitude of a spill if the chemical involved is present in an immiscible non-aqueous phase and has density significantly greater or less than water. The presence of a sinking organic phase poses obvious problems for contaminant recovery as well.

Summary

This paper has addressed the history, classification, and current issues concerning contamination, of groundwater by point and nonpoint pollution sources. Historically, pollution sources have resulted from a lack of understanding of the soil as a waste disposal system. This lack of understanding has manifested itself in the inadequacy of engineering designs and regulations to protect groundwater. Recent advances in our capability to detect pollution have resulted in widespread identification of contamination problems.

Sources of groundwater contamination arise from both point and nonpoint sources. Due to their nature, nonpoint sources are largely uncontrolled except through regulatory or incentive measures. The potential for pollution from point sources can be mitigated by proper facility siting, facility design and leak/leachate detection monitoring.

Control measures for nonpoint sources, and facility design and siting for point sources, should be based upon predictive modeling of the risk or disbenefits associated with impacts on the soil/groundwater system. Significant gaps exist in our knowledge to model such impacts. Major issues include the adequacy of existing model equation themselves to predict pollutant transport in soils, the effects of mixtures and cosolvents on the retardation phenomenon and the biodegradation of organic chemicals in soils and groundwater. Another significant drawback is the lack of field validation of existing models. Therefore, the understanding of how good (or bad) these models are is limited. Notwithstanding problems with model accuracy, the uncertainty in model predictions must also be established in order to properly perform risk assessments. This is a highly visible issue which is currently being addressed in both research and applications. Significant problems include both the paucity of data to characterize system variables and the in difficulty of accounting for this variability in modeling.

Controls for point sources, in addition to facility siting, include both prevention and detection measures. Some work has been done to assess the failure probabilities of components of facility designs but to our knowledge, this information has not been practically used in a risk management framework to evaluate the adequacy of designs. Detection systems for point sources have been utilized for only a relatively short period of time. Data to confirm the reliability of these systems is not readily available.

The current outcry in waste management practice is for "zero discharge" to groundwater. We have run the gamut over the past two decades from complacency to (perhaps) an extreme overreaction in dealing with this problem. Conservatism in regulation and design may be warranted at this stage. However, as our knowledge of soil and groundwater systems develops, we may find that such designs, which do not account for the natural ability of the soil and groundwater to purify themselves are overly restrictive and place an undue cost on society for disposing of wastes.

References

Besselievre, E.B.. 1969. The Treatment of Industrial Wastes. McGraw-Hill Book Company, NY.
Dean, J.D., A.M. Salhotra, E.W. Strecker, D.A. Grey and P.H. Howard. 1986. Aldicarb Exposure Assessment Florida: Further Simulation Studies. Draft Report for Contract No. 68-03-6304, U.S. Environmental Protection Agency, Athens, GA.
Fair, G.M. and J.C. Geyer. 1963. Water Supply and Wastewater Disposal. John Wiley and Sons, Inc., NY.
Federal Register. 1986. Environmental Protection Agency, Hazardous Waste Management System: Land Disposal Restrictions. Vol. 51, No. 9, January 14.
Freeze, R.A. and J.A. Cherry. 1979. Groundwater. Prentice Hall, Inc. Englewood Cliffs, NJ.
Gehm, H.W. 1976. Sludge Handling and Disposal. Chapter 17 in: Handbook of Water Resources and Pollution Control, H.W. Gehm and J.I. Bregman, eds. Van Nostrand Reinhold Company, NY.
Hagerty, D.J., J.L. Pavoni and J.E. Heer, Jr. 1973. Solid Waste Management. Van Nostrand Reinhold Company, NY.
Jury, W.A. 1985. Spatial Variability of Soil Physical Parameters in Solute Migration: A Critical Literature Review. EA-4228 Interim Report, Electric Power Research Institute, Palo Alto, CA.
Metcalf and Eddy. 1972. Wastewater Engineering. McGraw Hill Book Company, NY.
Novotny, V. and G. Chesters. 1981. Handbook of Nonpoint Source Pollution Sources and Management. Van Nostrand Reinhold Company, NY.
Overcash, M.R. and D. Pal. 1979. Design of Land Treatment Systems for Industrial Wastes - Theory and Practice. Ann Arbor Science Publishers, Inc. Ann Arbor, MI.
Overcash, M.R. and J.M. Davidson. 1980. Environmental Impact of Non Point Source Pollution. Ann Arbor Science Publishers, Inc. Ann Arbor, MI.
Shields, W.J., E.W. Strecker, J.D. Dean, and S.M. Brown. 1986. Chemical Spill Uncertainty Analysis. Draft Report for Contract RP2634-1, Electric Power Research Institute, Palo Alto, CA.
Todd, D.K. and D.E.O. McNulty. 1976. Polluted Groundwater: A Review of the Significant Literature. Water Information Center, Port Washington, NY.
Water Pollution Control Federation and the American Society of Civil Engineers. 1977. Wastewater Treatment Plant Design. WPCF, Washington, D.C.

SURVEY OF GROUNDWATER CONTAMINATION IN MASSACHUSETTS

Steven P. Roy*

Abstract

Public water supplies in Massachusetts have been contaminated or threatened from a number of sources. The purpose of this paper is to outline the nature of groundwater contamination and the mechanisms that the Department of Environmental Quality Engineering has employed to gather and assess information on groundwater quality threats.

One-third of the State's population obtain their water from groundwater sources. Public supply wells provide water to approximately 1.5 million people, and private wells provide water to an additional 400,000 individuals.

Groundwater contamination and drought aggravated problems have caused the loss of a number of groundwater sources. In the last ten (10) years 109 public water supply wells have been closed due to groundwater contamination. This represents a loss of approximately 50 million gallons per day. Forty-eight communities or 14% of the communities in the Commonwealth have experienced the loss of public water supply wells.

The identification of groundwater threats and contamination has been an on-going function of the Massachusetts Groundwater Protection Program. The major program activities that have contributed to this infromation base include: the Water Supply Protection Atlas; LeGrand ranking of waste sites; the Water Supply Contamination Correction Program; the Aquifer Land Acquisition Program; the Pesticide Evaluation Program; and the Community Technical Assistance Program. Each are reviewed.

Future activities will be in the areas of the development of a spatially referenced data base (geographic information system) and in the initiation of a water allocation permit program which will tie existing programs together for wellhead and resource protection.

*Groundwater Programs Manager, Commonwealth of Massachusetts, Department of Environmental Quality Engineering, Division of Water Supply, One Winter St., Boston, MA 02108

INTRODUCTION

It has become apparent in the years between 1970 and the present time that widespread contamination of groundwater has become the major environmental issue. Population increases and subsequent development pressures affect the availability of water supply lands during a time of increasing water supply demands. Water supply development and land use development is a local issue in Massachusetts which creates a major dilemma for local water supply agencies. Often they are placed into an adversarial position against other local officials that may be encouraging the siting of land uses that are incompatible with water supply protection. In the past, development interests often won the battle with the consequence being eventual contamination and loss of many public water supply sources.

Over the past five years the Massachusetts Legislature has passed innovative legislation that has allowed the D.E.Q.E. to develop a wide variety of programs to address the management, protection and cleanup of groundwater supplies. This paper will summarize the status of groundwater use, contamination and the programs that the agency has developed to respond to the management of our groundwater resources.

Geology

Virtually all of the State's groundwater withdrawals for municipal supply are from the stratified-drift aquifer. This is in contrast to the 400,000 private domestic wells which draw primarily from bedrock aquifers, except in southern Massachusetts, on Cape Cod, and on the Islands of Martha's Vineyard, and Nantucket where the only aquifer is stratified drift. The stratified glacial drift aquifers, which consist of sand and gravel with some silt, was deposited over bedrock by glacial meltwaters as the last continental glacier retreated from New England. Most of these deposits form small but very permeable valley-fill aquifers. These valley aquifers have a relatively small volume and storage capacity, however they are very productive because of induced infiltration from streams.

Most public water supply wells are sited close to these streams to take advantage of this situation. However, induced infiltration has depleted streamflows and some small streams cease flowing during major pumping periods.

The stratified drift outwash plain deposits in southeastern Massachusetts are much deeper and more continuous and serve as the major source of public water supply for the population in that area. Cape Cod and Nantucket Island have received Sole Source Aquifer designation from the EPA.

Table 1 presents the aquifer and well characteristics in Massachusetts.

Table I, Aquifer and well characteristics in Massachusetts (USGS, 1985)

Aquifer name and description	Well characteristics				Remarks
	Depth(ft)		Yield(gal/min)		
	Common range	May exceed	Common range	May exceed	
Stratified-drift aquifer; Sand and gravel with silt, glacial outwash, ice-contact, and delta deposits; some beach and dune deposits included. Moraines also contain till. Generally unconfined, locally confined.	60-120	200	100-1,000	2,000	Used extensively for public supply; also used for industry, fish hatcheries, agriculture, and rural supplies. Locally, large iron or manganese concentrations a problem. Some saline water intrusion in coastal areas. Low pH of water may corrode pipes and appliances.
Sedimentary bedrock aquifer: Red sandstone, shale, arkosic conglomerate, and basaltic lava flow. Generally unconfined, confined at depth.	100-250	500	10-100	500	Used for rural supplies and some industry. Deep wells produce hard water.
Carbonate rock aquifer: Limestone, dolomite, and marble. Confined.	100-300	1,000	1-50	1,000	Used for rural supplies and some industry. Water hard.
Crystaline bedrock aquifer: Metamorphic and igneous rock predominantly gneiss and schist. Confined.	100-400	1,000	1-20	300	Used for rural supplies. Locally, large iron concentrations a problem. Recently drilled wells generally deeper than older wells. Low pH of water may corrode pipes and appliances.

Water Supply Distribution in Massachusetts

Massachusetts has a very complex water supply distribution network. Three hundred sixty-three separate public water supply systems exist in 293 communities throughout the Commonwealth. The 363 systems consist of 68 private water companies, 78 water districts, and 217 municipal water departments; 58 communities are served entirely by private wells (Massachusetts Division of Water Resources, 1983).

Water supply is primarily a local responsibility in Massachusetts, with the major exception being the Massachusetts Water Resources Authority and five other regional systems which when combined supply fifty percent of the State's population with drinking water. "Although the State's major urban areas use surface-water supplies, groundwater is the primary source for 165 public supplies and a secondary source for an additional 33 public supplies (USGS, 1984). One-third of the 5.7 million people in Massachusetts obtain their water supply from groundwater.

SURVEY OF CONTAMINATION

The problem of groundwater contamination of public water supplies can be summarized as follows. As of October 1986, forty-eight (48) communities in Massachusetts have had public water supplies closed due to contamination. These closures represent the temporary or permanent loss of 110 public wells or wellfields across the State, or approximately a 55 mgd loss. The sources of contamination which resulted in the closures include: industrial discharges; accidental spills or leaks; road salting; landfill leachate; sewage treatment plants; agriculture and urban development.

Table 2. Sources of Public Water Supplies Closed Due to Contamination

Source	# Wells Closed
Industrial	18
Landfills	8
Road Salt	8
Fuel Storage	8
Agricultural	4
Illegal Disposal	3 ˈ
Overdevelopment	2
Sewage Treatment Plant	1
Natural Sources	18
Unknown	39
Total	109

To date, thirty-eight (38) communities continue to have closed public water supplies which in total includes 69 contaminated wells, 7 wellfields, and one reservoir. Since April 1975, eight (8) contaminated and closed wells have been brought back on-line after treatment. The rest remain closed or abandoned. Using an average per capita consumption rate of 70 gallons/person/day, and the loss of 55 MGD, approximately 786,000 people have been directly affected by this loss due to contamination.

GROUNDWATER MANAGEMENT PROGRAMS

In January 1983, the Department adopted a groundwater protection strategy with the stated goal of "preventing groundwaters and surface waters that recharge groundwater from being degraded or depleted, and to manage cases of known or suspected contamination to reduce the impact of contamination to levels consistent with the water's intended use, given public health, economic, and technical constraints". The 1983 Groundwater Protection Strategy set the organizational framework necessary to identify and effectively and efficiently address matters which affect groundwater quality and public health. Within the context of the Strategy, new regulatory, financial assistance and technical assistance programs have been developed. The following is a brief synopsis of these programmatic efforts.

Regulatory Guidance - Regulatory guidance is provided to cities and towns in Massachusetts through ongoing programs. Many of the regulations that DEQE develops and enforces protect groundwater quality. These include: sanitary landfill regulations, hazardous waste management regulations, wetlands protection regulations, on-site and municipal wastewater treatment regulations, drinking water regulations, and groundwater discharge permitting and classification.

Technical Assistance - Another major element of the program is technical assistance and information exchange. Numerous handbooks for local officials have been developed, a bimonthly newsletter is produced and educational and technical workshops are presented. One of the key components of technical assistance is the water supply Protection Atlas. The Atlas consists of four overlays for each of the 177 U.S.G.S. 7.5 minute topographic quadrangles covering Massachusetts. Waste sources, water supplies, drainage basins and aquifers are depicted on these overlays. The Atlas serves as a visual representation of the relationship between existing or planned water supplies to known or suspected sources of groundwater contamination. Each of the 1600 waste sites have been evaluated for their potential threat to nearby public water supplies using the LeGrand (1980) system. The waste source and water supply overlays have been digitized for eventual use in a statewide geographic information system. Direct assistance to cities and towns on landuse planning and local regulatory controls for groundwater protection is also provided.

Financial Support - A key component to the Massachusetts groundwater protection program is economic assistance to communities to strengthen their ability to protect, manage and develop water supplies. Funds are provided to communities for immediate response to water supply emergencies and remedial cleanup of contaminated water supplies. Fifteen million dollars are available under the Water Supply Contamination Correction Program for this effort. Currently, eighteen projects are underway to restore public water supplies lost due to contamination. In addition, this program supports a major departmental effort in the area of widespread agricultural contamination. Groundwater contamination by ethlyenedibromide (EDB), Temik, Dichloropropane and other pesticides have threatened and closed both public and private water supplies. A major mapping project is now underway to correlate areas of known crop types to potential hotspots for groundwater contamination. Through the use of Geographic Information System (GIS) technology, multiple overlays of information can be prepared, analyzed and visually displayed.

Another major aspect of financial assistance in the program is the Aquifer Land Acquisition Program (ALA). The ALA program provides funding to communities to purchase land over sensitive recharge areas. $14 million dollars is available under the program to protect sensitive recharge areas around water supplies of high quality and quantity through a combination of strategic land acquisition and effective land use controls. Twenty-six projects are currently underway whereby recharge areas are being defined, lands purchased and landuse controls established. Approximately 5000 acres will be purchased in fee simple under the existing funding allocation. An additional 3000 acres will be controlled through the use of restrictive easements, and approximately thirty-five hydrogeologic studies will be funded to define recharge areas. Maximum grant awards are limited to $500,000 per project. Dozens of local land use controls have also been implemented as a result of this program.

These economic incentives for groundwater protection, management and cleanup should assist in strengthening the ability of local authorities in protecting, conserving, and developing their groundwater supplies. By encouraging local responsibility, and long-range water supply planning, this funding will assist local governments in effectively managing their water supplies.

Future Programs - Work is currently underway to begin implementation of the Massachusetts Water Management Act. DEQE is granted broad regulatory authority to oversee all usage of water within the Commonwealth, including the prioritization of applications for withdrawal permits. All withdrawals in excess of 100,000 gallons per day will require a discharge permit. The program will mandate proper water supply planning and management for all sources of water supply in the Commonwealth.

SUMMARY

The groundwater protection efforts within Massachusetts are guided by the goal, principals, and programs as defined in the Massachusetts Groundwater Protection Strategy. Numerous contamination incidents have prompted the state to take an active role in the cleanup and protection of groundwater wells. Financial support in cleanup and protection of water to municipalities provides the major stimulus for proper groundwater management and protection.

References

U.S. Geological Survey. 1985. National Water Summary 1984. Water Supply Paper 2275. 467p.

Massachusetts Division of Water Resources. 1983. Massachusetts water supply, safe yield, type of supply, proposed sources: Boston, 47p.

LeGrand, H.E. 1980. A Standardized System for Evaluating Waste-Disposal Sites. National Water Well Association. 42p.

SYNTHETIC ORGANIC CONTAMINANTS AND PESTICIDES IN GROUNDWATER

Mahfouz H. Zaki, M.D., Dr.P.H.*

For decades, the groundwater of Suffolk County, New York, has been bacteriologically and chemically of superior quality despite the high iron and manganese occasionally encountered in a few areas. In late 1976, the Suffolk County Department of Health Services became aware of the identification of a few synthetic organic compounds in wells in neighboring Nassau County. As a result, the Department initiated an extensive monitoring program to determine the presence of selected halogenated hydrocarbons in groundwater and, if so, the extent of the problem.

Synthetic Organic Compounds and Petroleum Products

When first initiated, the program concentrated on community water systems, namely municipal water supplies and private water purveyors which serve year-round residents, but was later expanded to cover non-community water systems and private wells.

To date, more than 39,000 samples from community and non-community water systems and private wells have been examined for the most commonly used halogenated hydrocarbons, such as trichloroethylene, tetrachloroethylene, 1,1,1 trichloroethane, trichlorotrifluroethane and a few others.

Water samples were considered unacceptable if they exceeded the guideline recommended by the New York State Department of Health, namely, 50 ppb for individual ingredients and 100 ppb for all contaminants. For benzene and vinyl chloride, the acceptable concentrations were 5 ppb.

Of all community water systems tested, about 3.3 percent exceeded the guideline of 50 ppb for individual compounds or 100 parts for the aggregate. Thirteen percent showed traces below the guidelines. Of the non-community water systems and private wells, about 3.9 percent exceeded the guideline and 20.0 percent showed traces below the guidelines. The following table shows the results of testing of public and private water wells for the three most commonly encountered synthetic organic compounds:

*Director, Division of Public Health, Suffolk County Department of Health Services, 225 Rabro Drive East, hauppauge, N.Y. 11788

Results of Testing For Selected Synthetic Organic Compounds In
Public and Private Water Supplies in Suffolk County, New York

Compound	Percent exceeding 50 ppb	Percent with traces less than 50 ppb	Mean concentration in ppb
Trichloroethylene			
Public	0.6	5.9	45
Private	1.1	4.5	183
Tetrachloroethylene			
Public	0.5	5.2	28
Private	1.2	6.6	155
1,1,1 Trichloroethane			
Public	0.5	13.0	14
Private	3.0	18.9	113

Irrespective of the validity of the actionable levels which have been utilized since 1976, Suffolk County, as a responsible health agency had no alternative but to assume that drinking water containing the organic contaminants above the recommended levels is hazardous and to act accordingly.

In the case of community water systems, the purveyors were advised to refrain from using the contaminated well and to switch to alternate uncontaminated ones. Purveyors were able to comply because of the presence of alternate uncontaminated wells in the same well fields.

In the case of non-community water systems and private wells, the owner or manager was advised to either:

a. Connect to public water mains, if accessible and fiscally feasible
b. Deepen the well
c. Relocate the well
d. Use an activated carbon treatment unit

In a few situations, the identification of a cluster of contaminated wells and the diversity of contaminants in high concentrations were indicative of industrial discharges and/or accidental spills. In an effort to locate the possible source or sources of contamination, industrial surveys were undertaken. Investigations revealed that, in a few instances, illegal industrial discharges or spills might have occurred. Legal action against some industrial concerns was initiated.

Pesticides

Apart from synthetic organic compounds and petroleum products, the County got heavily involved in the investigation of groundwater contamination with pesticides.

As you know, the use of pesticides is regulated by the federal government through the Environmental Protection Agency and the corresponding state agencies. Permits are issued after extensive laboratory and field investigations demonstrate that the pesticide in question does not have significant deleterious effects on human health, the nontarget populations or the environment.

In spite of all these prerequisites, rules and regulations required for the issuance of a permit, the existence of a combination of circumstances can allow a pesticide to reach the groundwater in concentrations which may be potentially hazardous to human health.

On August 24, 1979, the United States Environmental Protection Agency informed the Suffolk County Department of Health Services that water samples collected from a few wells in Eastern Suffolk County contained traces of a pesticide, aldicarb, and that additional water samples would be collected for confirmatory purposes.

Aldicarb is a highly toxic carbamate pesticide manufactured by the Union Carbide Corporation under the trade name of Temik. It is a member of a group of pesticides which exert their pesticidal activity through inhibition of cholinesterase. It is metabolized to aldicarb sulfoxide and finally to aldicarb sulfone, both of which are also pesticidal. The parent compound is active against several insects and some nematodes by systemic action and so are its metabolites.

Temik is registered in the United States for use in cotton, potatoes, peanuts, sugar cane, sweet potatoes, sugar beets and ornamental plants. The doses and mode of application vary in the different crops.

In spite of its high toxicity, accidental poisoning is rather infrequent. During 1977, 9 individuals who ate cucumbers contaminated with 8,000 - 10,000 ppb of aldicarb experienced one or more of the following symptoms: Nausea, vomiting, blurred vision, dyspnea, perspiration, headache, and temporary paralysis of extremities which lasted only 4-12 hours with no residual effects. Another incident reported recently involved the consumption of melons which had a high concentration of aldicarb.

In order to find whether any cases of carbamate poisoning have been reported in the County, the Department of Health Services contacted the local hospitals and poison control centers to inquire about their receiving or treating any cases of carbamate poisoning during the past few years. Our investigation revealed that no cases were known to these agencies.

Testing of a few hundred wells by EPA and the State Laboratory indicated that the pesticide was present in varying concentrations in groundwater in several areas in Eastern Suffolk County.

Because of its universal use in the estimated 22,000 acres of potato farms and because of Suffolk County's unique position as an area with a sole source aquifer, an extensive monitoring program was initiated by the Suffolk County Department of Health Services in cooperation with the Union Carbide Corporation.

A mass survey was conducted covering an area of approximately 100 square miles. The area was divided into grids 1,500 x 1,500 feet. All accessible wells within 2,500 feet of potato farms were sampled. During an 8-week period, more than 8,000 water samples were collected and were shipped to the Union Carbide laboratories in Charleston, West Virginia for testing. Split and spike samples were sent to Union Carbide and other laboratories for confirmatory purposes.

The acceptable level which was utilized in the survey was the one recommended by the National Academy of Sciences and which was adopted by the State Health Department. The Academy considered the presence of seven parts per billion of aldicarb to be safe for human consumption. Several factors were considered and assumptions made in calculating this no-adverse-effect level in drinking water. These included:

o   That the no-adverse-effect dose-based on animal and human studies-is 0.1 mg/kg body wt/day;
o   That when an uncertainty factor of 100 is utilized, the acceptable daily intake would be reduced to 0.001 mg/kg/day;
o   That the average adult weighs 70 kilograms;
o   That the average adult drinks two liters of water per day; and
o   That the aldicarb intake from water represents only 20 percent of the total aldicarb intake.

This 7 ppb level has been challenged by Union Carbide which questioned the validity of some of the aforementioned assumptions. Several issues were raised including:

o   That the EPA acceptable daily intake of aldicarb is .003 mg/kg body weight;
o   That the no-adverse-effect level in animals is 0.125 mg/kg body weight;
o   That a tenfold margin of safety has been recommended by EPA in situations where cholinesterase inhibition is the major and/or only effect noted in animal studies; and
o   That limiting the aldicarb intake from aqueous sources to 20 percent cannot be substantiated on any grounds.

Several no-adverse-effect levels in groundwater were consequently proposed using different methods of computation and acceptable daily intake. These ranged from 21 ppb to 100 ppb.

The question of whether deleterious health effects can or do occur at this very low level had to be addressed. Unfortunately, hardly anything is known about the long-term health effects resulting from the consumption of trace concentrations of aldicarb. Again we had no alternative but to assume that concentrations of aldicarb exceeding 7 ppb are potentially hazardous and to act accordingly.

The following table shows the results of testing of private wells, non-community water supplies and community water systems during and prior to the mass survey. (Zaki et al, 1982):

Aldicarb Concentrations in Groundwater by Type of Well

Source	Number of wells	Exceeding the Guideline	Showing Traces	Non-detectable
Private Wells	8051	1087 (13.5)	1068 (13.3)	5896 (73.2)
Non-Community Water Supplies	274	22 (8.0)	45 (16.4)	207 (75.5)
Community Water Supplies	68	5 (7.4)	5 (7.4)	58 (85.3)
Others	11	7	1	3
Total	8404	1121 (13.3)	119 (13.3)	6164 (73.3)

Figures in parentheses indicate the percentages.

As is evident, 13.5 percent of the private wells exceeded the guideline, while 8 percent of the non-community
water supplies and 7.4 percent of the community water systems did so.

Of the 8,051 private wells tested, marked variation
occurred in the extent of contamination in different areas
within the same townships and within 2,500 feet of potato
farms. This variation is attributed to the hydrology of
the area, the soil conditions, the movement of groundwater,
and the dose and time of application of the pesticide.

Health Effects

Exposure to high doses of certain synthetic organic
compounds and pesticides, as in the case of accidental
industrial exposure, affects the central nervous system
resulting in depression, headache, blurred vision, nausea,
vomiting, convulsions, loss of consciousness and in some
instances, death. Some compounds, such as carbon tetrachloride or chloroform may cause hepatotoxicity or renal
failure. In animals, some of these compounds were found
to cause cancer of specific organs.

In the early seventies, the National Cancer Institute
demonstrated an increase in liver and kidney tumors among
animals exposed to high doses of chloroform. During the
same period, attention was focused on the association
between halogenated hydrocarbons in drinking water and the
incidence of human carcinoma.

Reports published by Harris and Page showed a slightly
higher mortality rate from cancer of the urinary bladder
among residents using the water of the Mississippi River
than among those using ground water. The assumption was
made that the river water contained more organic matter
than groundwater and that the former was usually chlorinated.

Other studies conducted during the same decade in Ohio,
Upstate New York, Maryland and Los Angeles showed an association between chlorinated water or chloroform in drinking
water and certain cancers.

These reports prompted the Environmental Protection
Agency to ask the National Academy of Sciences to have a
group of investigators study the carcinogenic effects of
chloroform and other trihalomethanes in drinking water.
The investigative team conducted an extensive review of
epidemiologic studies, published and unpublished.

Most of the studies, however, used indirect evidence
of the presence of trihalomethanes in drinking water. This
meant that comparisons of cancer rates were made between
populations receiving surface waters and those using groundwaters. Very few studies had actually quantitative measurements of the trihalomethanes in water. The panel concluded
their comparison by stating that:

"None of the studies reviewed were able to adequately
take into account many well established risk factors
for cancer rates at different sites. For example,
for bladder cancer, one needs to control for occu-
pation, cigarette smoking, use of alcohol and drugs,
non-aqueous sources of THM, coffee consumption,
socioeconomic status and ethnicity"...

"The results do not establish causality, and the quantitative estimates of increased or decreased risk are
extremely crude. The effects of certain potentially
important confounding facts, such as cigarette
smoking, have not been determined." (National Academy
of Sciences, 1978)

In order to assess the health effects of long, contin-
ued exposure to low doses of carcinogenic substances, the
Safe Drinking Water Committee of the National Research
Council adopted the following principles:

1. That the effects in animals, properly qualified,
   are applicable to man.
2. That methods to establish a threshold for long-
   term effects of toxic agents are unavailable.
3. That exposing experimental animals to high doses
   of toxic agents in order to determine the carcinogenic effects is valid and justifiable.
4. That materials should be assessed in terms of
   human "risk", rather than as "safe" or "unsafe".

Accepting these principles as a framework, the
Committee reviewed innumerable studies in terms of methodology and statistical models used, discussed their limitations, and made several recommendations which are included
in the most comprehensive report entitled Drinking Water
and Health, by the National Academy of Sciences (1977).

For the sake of illustration, human cancer risk associated with trichloroethylene in drinking water was based
on experiments in which male and female mice were fed corn
oil containing trichloroethylene by a tube for 78 weeks.
Surviving mice were sacrificed at 90 weeks and examined
microscopically.

Risk estimates were expressed as the probability of cancer after a lifetime consumption of 1 liter of water/day containing Q ppb of the compound.

A risk of 1×10^{-6} Q implies a lifetime probability of 1 cancer case per million if the concentration of carcinogen was 1 ppb and those exposed drank 1 liter per day.

When 2 liters are consumed (which is considered by the Committee to represent the average daily intake) and if the concentration of the carcinogen was 10 ppb, the lifetime probability would be 2×10^{-5} or 2 cases per 100,000 persons.

The tremendous achievements in laboratory technology during the past decade have enabled us to detect traces of many contaminants previously unidentified. Although the toxic effects of many contaminants are well documented in heavy industrial and acute exposure, very little is known about the long-term effects of exposure to low doses of these contaminants.

In order to establish actionable levels for many synthetic organic compounds and pesticides, reliance has been placed on extrapolation from animal experiments and on the few uncontrolled epidemiologic studies among human populations. In the process of determining these levels, several parameters were used including the no-adverse-effect level, acceptable daily intake, water consumption, expected exposure to other non-aqueous sources, etc. The limitations inherent in extrapolation from animal studies and from the crude uncontrolled epidemiologic studies, and the many stipulations and assumptions utilized, subject these proposed actionable levels to serious questions and challenges.

Probably the most important task at present is a reevaluation of the actionable levels of the frequently encountered environmental contaminants. This reevaluation should take into consideration realistic health effects, reasonable acceptable risks and economic impact.

One of the main obstacles in establishing reasonable guidelines is the lack of controlled epidemiologic studies. It is fully realized that it might be very difficult and perhaps impossible to design a study which could establish a cause and effect relationship between synthetic organic compounds and pesticides and the clinical manifestations characteristic of these contaminants because of the various etiologic agents which could be incriminated, the variations in the degree of exposure to contaminants in water and other non-aqueous sources, and the long follow-up

periods. Despite these anticipated difficulties, it is
of paramount importance that studies be initiated to
explore at least the presence or lack of association between pesticides and selected clinical conditions.

The lack of evidence to support or negate a causal
relationship at present does not mean that we, as public
health officials, can ignore the presence of these toxic
contaminants in groundwater until further documented studies
prove the presence or absence of such evidence. On the
contrary, any responsible health agency should assume that
these contaminants represent a potential hazard to public
health and act accordingly.

In Conclusion:

The past decade has witnessed a dramatic emphasis on
environmental contaminants and their impact on human health.
The wide publicity given incidents such as the Love Canal
in New York State and the possible deleterious effects of
exposure to Agent Orange have heightened public awareness
and, in a sense, engendered a state of environmental
paranoia.

We should do everything possible to identify and abate
sources of pollution and find corrective solutions to
existing sources whose effects are tangible. It would be
most desirable, of course, to avoid exposure to all possible
risks - the tangibles and intangibles. This requires
tremendous resources which are not, and most probably will
not, be available in the near future.

Our society seems to be willing to accept tangible and
measurable risks in our daily activities as a result of
cigarette smoking, excessive food and alcohol intake, and
the use of the automobile. The same society, however, is
unable to tolerate potential, intangible, and unmeasurable
risks from food additives, pesticides, air pollutants, and
water contaminants. As public Health administrators, we
have a responsiblity, not only to monitor and control
these substances, but also to help the public become fully
aware of their risks and benefits without resorting to
rhetoric on either side of the issue.

References

Zaki, M.H., Moran, D. and Harris D., Pesticides in
Groundwater: The Aldicarb Story in Suffolk County,
New York, American Journal of Public Health, 72,
1391-1395, 1982

National Academy of Sciences, Epidemiology Subcommittee
of the Safe Drinking Water Committee: Epidemiological
Studies of Cancer Frequency and Certain Organic
Constituents of Drinking Water. A report submitted
in 1978.

National Academy of Sciences, Safe Drinking Water
Committee: Drinking Water and Health, 1977

AN OVERVIEW OF GROUND WATER MONITORING TECHNIQUES

R. W. Schowengerdt*

ABSTRACT

The transport of contaminants in a given ground water regime is a function of three interrelated, complex factors: contaminant concentration dynamics, system hydraulics, and geologic control. Ground water investigations designed to determine the extent and nature of contamination with the ultimate goal of site remediation rely on the accurate definition of these hydrogeologic parameters. Numerous methods are currently being utilized to characterize ground water conditions associated with hazardous waste sites ranging from extremely sophisticated installations and instrumentation to relatively simple field observations. The goal of this paper is to review a cross-section of ground water monitoring techniques and methodologies in practice today. Major emphasis is placed on high integrity monitoring required at hazardous waste sites as mandated by regulatory authorities and legal liability.

INTRODUCTION

Ground water contamination resulting from the disposal of hazardous chemicals, governed by the present regulatory climate and increased public awareness, has served to focus on the techniques used to evaluate subsurface hydrologic regimes. Numerous methods are currently being utilized to characterize ground water conditions associated with hazardous waste sites ranging from extremely sophisticated installations and instrumentation to relatively simple field observations. The intensity of these efforts are directly related to liability and associated remediation costs. The capital expenditures required to mitigate a site to an acceptable risk level typically range into the tens-of-millions of dollars. With final remediation costs of this magnitude, the behavior of the hydrologic system must be quantified. Ground water investigations designed to determine the extent and nature of contamination with the ultimate goal of site remediation rely on the accurate, defensible definition of the hydraulics and chemistry of the system.

The majority of ground water studies associated with hazardous waste occurrences involve a contaminant outbreak to a visible source such as

*Project Engineer/Hydrologist, International Technology Corporation
11270 West Park Place, Suite 700, Milwaukee, Wisconsin 53224

a municipal well, a river, or a leachate spring. At this point, the project engineer is faced with quantifying the contaminants, determining the source, and defining the ground water hydraulics to an extent which will allow remedial design alternatives. This paper presents a progressive, step-by-step process of the activities which are typically employed by the author to characterize the ground water regime at a hazardous waste site. Each step in the process systematically builds the data base and further directs subsequent activities to "fine-tune" the information output. This approach minimizes the collection of unnecessary data and continually refocuses the later more detailed assessments to answer specific questions and fill data gaps.

A cross section of investigation techniques are presented under each of the progressive steps as an overview. In the final design of a ground water investigation it is up to the project engineer to decide which group of techniques will be employed based on client needs, site specific conditions, liability factors, and budget constraints.

DOCUMENTATION

Prior to the discussion of ground water investigation techniques, a short preview of data validity and quality assurance/quality control (QA/QC) is warranted. In light of the liability associated with ground water contamination, the verification of data is essential. A quality assurance project plan (QAPP) which details testing methodologies, data verification, and field procedures must be prepared and followed. The QAPP provides the foundation for data accuracy, precision, and legal defense, if required. The QAPP also provides a basis which enables all data, calculations, laboratory analyses, and field testing procedures to be independently duplicated. Specific elements contained in a typical QAPP range from verification of data collected by others, to laboratory QA/QC procedures such as spike, duplicate, and surrogate analyses, to instrument calibration, and calculation checkprints. The United States Environmental Protection Agency (U.S. EPA, 1980) has developed a set of guidelines to be considered when preparing a QAPP. Table 1 presents the table of contents of a typical QAPP. The QAPP ultimately directs all investigation procedures to standard and consistent methodologies resulting in a defensible product.

Written documentation of field procedures, observed conditions, field data, sampling locations, times, instrument calibration, decontamination methods, sample preparation, or any other pertinent site information is required to justify the final interpretations. Most consulting firms and laboratories have their own forms which must accompany all laboratory samples (such as chain-of-custody records) and field data (boring logs) entries. Without this legal record, the data may be rendered useless in the event of a third party challenge (which is commonplace in hazardous waste situations). Written documentation entered during the actual field work (not completed after the fact) cannot be under emphasized by the prudent project engineer.

40 CONTAMINATED GROUND WATER

EXISTING DATA - PHASE I

The first step in designing a ground water investigation is to define the problem and conceptualize how it can be solved. In the hazardous waste instance, the problem often is highly visible such as a leachate seep or a contaminated well. Many times the state and/or federal regulatory authorities have already become involved and have mandated a response. An assessment of the information available will formulate a preliminary response. Generally, three types of existing information are available to initially characterize a site: (1) Site specific analytical information, (2) Regional geologic-hydrologic reports, and (3) Industrial records. A review of regulatory authorities' files will often reveal laboratory testing results and site visit notes which bracket the problem. As a response to the regulatory mandate, the private parties involved will often test the problem area independently. Other consultants may also have conducted site specific studies which may be useful in designing a comprehensive assessment. As part of the existing information assessment, all analytical data must be verified as per procedures outlined in the QAPP.

Regional geologic and hydrologic reports prepared by state and federal entities (such as the USGS and State Geologic Agencies) are usually available. These information sources can target specific water bearing strata of importance and provide a geohydrologic framework from which site specific investigations can be focused.

Industrial records vary from detailed disposal quantities and content records to employee recollections of the process related to the industry. The composition of the raw materials, intermediate components, and final product indicate potential contaminants. Industrial synthesis processes and disposal practices also will further delineate the likelihood of off-site migration of hazardous materials.

Other existing information may be available through a number of other sources, such as landfill reports, aerial photography, related environmental and engineering studies, remote imagery, and local planning documents.

Once all available information has been collected, verified and evaluated, the problem can be targeted and a ground water investigation can be conceptualized. The investigation design will begin with well defined field activities which will determine the specific details of later work.

SURFICIAL INVESTIGATIVE METHODS - PHASE II

Although ground water cannot be seen on the earth's surface, a variety of surface techniques such as geophysics and vapor surveys have potential to provide information pertaining to ground water occurrence and quality characteristics. Surface investigations do not provide a complete hydrogeologic picture; however, these methods are significantly less costly and do provide a reconnaissance level site specific definition to guide later activities. Surface techniques may also function to further delineate contaminant boundaries, such as buried landfill limits, drum locations, or a hydrocarbon plume. Remote sensing from aircraft or satellites has also become an increasingly valuable tool for understanding ground water conditions.

A variety of surficial geophysical techniques are currently employed to measure geologic, hydrologic and contaminant characteristics at hazardous waste sites. Geophysics applications measure differences or anomalies of electrical and magnetic properties within the earth's crust. Differences in these properties correspond to geologic structure, porosity, rock type, water content, water quality, and unnatural disturbances.

Geophysical methods often lack detailed resolution and need to be supplemented by later subsurface investigations. Results provided by geophysical testing, however, allow for extensive spatial coverage at a reasonable cost. Four of the most common geophysical methods are geomagnetics, electromagnetics, resistivity, and ground probing radar. These techniques exhibit varying degrees of complexity and only a brief synopsis is warranted herein. Griffiths and King (1965) present a detailed description of applied geophysical techniques.

Geomagnetics (GM) involve the measurement of a magnetic field induced by site specific conditions. This method is commonly used when metallic objects are thought to be buried (waste drums) and when the geologic materials have been altered as a result of natural geologic phenomena (fault areas) or by man-caused alterations (burial of non-native materials in a landfill). Typically, a grid of measurement stations is located throughout the study area. The field crew proceeds from station to station taking magnetic readings with a portable GM instrument at each station. The readings can then be plotted and contoured to illustrate geologic boundaries and/or buried metallics.

Electromagnetics (EM) measures the conductivity differences between a given distance of geologic material. The EM equipment typically used consists of a transmitter and receiver coils separated by about 12 feet. The portable unit is operated by one man and is most sensitive to near surface readings. EM surveys are particularly useful in contrasting very conductive areas or solutions from nonconductive regions. Boundaries of landfills saturated with leachate are accurately located using EM methods. The geometry of conductive contaminant plumes in ground water also can be readily detected. Field procedures are similar to those used in GM resulting in a contour map of iso-conductivities.

Electrical resistivity (ER) geophysical methods are similar to EM techniques in that electric resistivity (lack of conductivity) is measured between groups of electrodes placed in the ground surface. ER surveys are useful in determining lithologic unit boundaries (aquifer limits), porosity, and fluid composition. ER methods can provide deeper interpretations than conductivity although the field procedures involving ER are more rigorous. Applications of ER are restricted to conditions which allow an adequate electrical ground connection (warm climate and moist surficial materials). The field measurements in ER surveys can be positioned as a grid-point type as previously mentioned or arranged in a linear pattern to provide a cross section.

Ground probing radar (GPR) utilizes an FM-frequency pulse source, antenna, and a graphic recorder to map reflections from subsurface interfaces. The depth of the penetration of the radar energy is a function of the nature of the material. The difference in energy impedance can then be compared to known saturated and unsaturated lithologic responses (lesser penetration - saturated clays, greater penetration - sandy soils). The stratified sequence of undisturbed soils is easily distinguished from a disturbed randomly filled area (landfill or buried waste lagoon).

Since most geophysical techniques utilize variations in electromagnetic signals, background interference sources such as power lines or pipelines must be considered in the planning stages and subsequent analysis.

With the increasing occurrence of petroleum and organic solvent leaks, the use of vapor surveys has become widely used. When volatile organic compounds are involved in ground water contamination, dissolved or floating product plumes occur. A component of these contaminants are present as a vapor or soil gas phase which indicates the plume's spacial extent. Portable photoionization meters or "sniffers" can detect these vapors through shallow boreholes. As with the geophysical methods, a sampling grid is marked in the study

area.   A borehole is drilled to within 1-2 feet of the plume and
immediately "sniffed". The peak concentration is recorded at each
borehole (sampling point) to be contoured later. Vapor surveys have
proven to be especially effective in delineating gasoline and diesel
fuel plumes, allowing recovery wells to be located in the most cost
effective locations. A similar approach can be utilized to detect the
limits of buried landfills by evaluating methane gas occurrence.
Figure 1 presents an areal view of a landfill as defined by surface
geophysical, vapor survey and remote imagery techniques.   Several
photoionizers are currently available and care must be taken to assure
the meter is sensitive to the vapor in question (for example, some
"sniffers" do not detect methane).

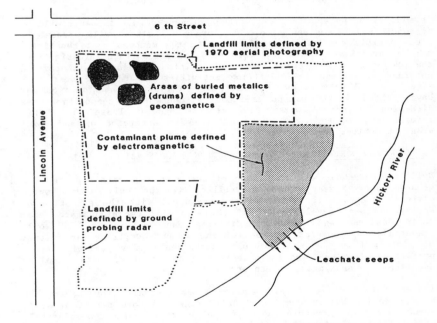

FIGURE 1 LANDFILL CHARACTERIZATION USING SURFICIAL METHODS

Remote imagery and aerial photography can provide useful information
regarding ground water conditions. The ready availability of photographs from government agencies and commercial activities has stimulated their use.   Observable patterns, colors, and relief make it
possible to distinguish variations in geology, soils, moisture
profiles, vegetation, and land use.   Color infrared images readily
characterize springs and seeps representing discharge areas.   Plumes
of leachate seeps discharged into native water bodies may also be
apparent in remote imagery.   Temperature sensitive films can dis-

tinguish hydrologic cycling in some areas. Lillesand and Kiefer
(1979) provide comprehensive documentation of remote sensing
applications.

The observation of geohydrologic features on and adjacent to a study
area can lead to determination of ground water flow patterns. The
location and elevation of springs, creeks, swampy areas and other
surface waters often partially define the ground water system and its
direction of flow. Observations of ridges, surficial geologic
changes, and other topographical characteristics all may add to the
understanding of the geohydrologic regime. Knowledge of depositional
and erosional environments in the study area may indicate the extent
and water level fluctuation of water bearing sequences.

It must be noted that not all methods presented in this paper will be
appropriate for a given site. The prudent engineer will assess the
site specific conditions and incorporate the methods most beneficial
to the investigation. At the completion of the surface investigation,
the project engineer should be in a position to use the information
obtained to direct well drilling activities towards the most cost
effective and data oriented locations. A typical investigation may
have yielded the contaminant source area, contaminant composition, and
direction of ground water flow at the end of Phase II. This
information, in turn, would be utilized to determine up and downgradient monitor well locations.

DRILLING AND MONITORING WELL CONSTRUCTION - PHASE III

Drilling procedures are often one of the most expensive aspects of a
ground water investigation. Drilling is the only method which
directly accesses the ground water resource, with the exception of
spring and seep development. If the surficial techniques are carefully designed and executed, drilling can be minimized to only
essential, supplemental portions of the study. Drilling can further
be minimized by the multiple use of boreholes for geologic interpretations (coring), aquifer testing (well tests), water quality
sampling, and as a product recovery well.

Investigations of relatively shallow sites typically incorporate the
use of mobile hollow-stem auger drill rigs. A core barrel or split
spoon sampler is advanced ahead of the augers to collect a relatively
undisturbed geologic sample (Acker, 1974). These samples are analyzed
in the field for structural, engineering, hydrologic, and lithologic
characteristics. Portions of the cores may be preserved for later
laboratory analysis of geotechnical or chemical properties. Field
supervision of the drill rig and crew is extremely important with
final field decisions (such as drilling depth, speed, or use of
additives) often affecting all subsequent data. Once the final
drilling depth is determined, a monitoring well is generally constructed inside the auger flights. Well completion decisions again
may effect outcome of the investigation and require an experienced
field engineer. Based on the properties observed during coring, the
specific well design is determined in the field. Recalling the
liability issues mentioned earlier and the need for representative

data, well construction must provide the integrity required for the
project. The open interval (slotted or screened section) of the well
should correspond exactly to the zone to be tested and monitored.
Wells completed across multiple hydrostratigraphic units are virtually
useless. Thousands of wells have been installed which lack this
integrity and make data analysis difficult and indefensible. If a
contaminated zone occurs above the aquifer to be tested, the borehole
needs to be cased and isolated (cemented off) prior to drilling into
the uncontaminated zone. The well screen or slotted casing and the
blank casing must be nonreactive with the suspected contaminants.
Stainless steel, Teflon®, and Bisphenol-A Epoxy represent examples of
relatively nonreactive casing; whereas, PVC casing exhibits sorbent
and leaching characteristics with many organic constituents. The
reactivity of these casing materials are available from the individual
manufacturers and should be consulted when selecting well materials.
The slotted or screened portion of the casing should be sand packed
with a nonreactive appropriately sized aggregate with both ends sealed
with bentonite (Driscoll, 1986). The remaining annular space should
be cemented with a bentonite/clay mixture to the surface. Finally,
all monitoring and test wells must be installed with a locking steel
surface casing to minimize well disturbance. Following well completion, each well must be developed to establish the sand pack and
clean the well bore allowing fresh formation water to be produced.
Well development procedures rely on induced movement of water through
the slots, screen and formation. Driscoll (1986) describes the
details of several development techniques in use today. During
drilling, well construction, and development operations, care must be
taken to prevent well contamination. Steam cleaning or other
disinfecting procedures must be conducted as prescribed in the QAPP.
The use of drilling fluid additives and lubricating oil and greases is
generally prohibited. During development, only bailing, pumping with
non-contaminating pumps, or air lifting with non-oil aerosol com-
pressors is acceptable. One other important consideration is that all
developed water, drill rig wash-down fluids and decontamination
rinsates may have to be collected and disposed as hazardous waste.

The drilling of deeper holes may require the use of larger rotary or
cable tool rigs but the general drilling and well completion procedures are similar to those summarized for shallow holes.

Many types of well completions are in use today to obtain water
quality samples, measure water levels, and conduct aquifer testing.
Well completions range from the standard cased monitoring well of
numerous diameters to pneumatically pushed piezometers, and driven
well points. Specialty type completions include ultra-small, pressure
driven sampling systems (which basically function similar to a beer
keg), to multiple point sampling systems, in which one completion can
individually monitor many zones. Figure 2 illustrates a typical
monitoring well and pressure driven system. Wells are many times
installed in close proximity to each other but monitor different
depths (nested piezometers). These completions are especially useful
in leakage evaluations and water level differentiations.

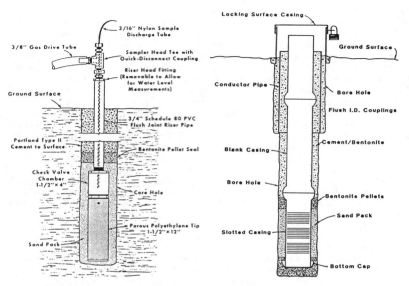

FIGURE 2 TYPICAL MONITORING WELL AND PRESSURE SYSTEM INSTALLATIONS

A vast number of installation techniques, materials and drilling methods are available today. It is the project engineer's responsibility to match the variables to the project needs, data integrity, and budget constraints. The construction of monitoring wells has only provided the conduits required to access the ground water resource for sampling and testing. All subsequent testing and the ultimate project success relies on the quality and location strategy of the monitoring well completions.

AQUIFER HYDRAULICS - PHASE IV

To make determinations of contaminant movement and design remedial measures (pumping well fields), the hydraulics of the ground water system must be characterized. Once the governing hydraulic parameters are quantified, several mitigation options can be simulated and evaluated for effectiveness.

Based on the data collected during the coring process and guided by the surficial investigations, aquifer testing may be conducted in the boreholes and/or monitoring wells previously completed. The aquifer testing program must be designed to provide the information required to define the hydraulics of the site specific system. The topic of aquifer testing is exhaustive and the techniques discussed herein are intended as an overview focusing on hazardous waste evaluations. The interested reading is directed to Kruseman and DeRidder (1983) for further aquifer testing details. In general, four types of data are required to characterize the ground water flow system: (1) Hydraulic conductivity, (2) Storativity, (3) Presence of leakage, and (4) Head distribution.

Head distribution is the measure of the water level elevation in a monitoring well. Single well tests in which the system is perturbed by pumping, bailing, injection, gas pressuring, or volume displacement, provide estimates of hydraulic conductivity (essentially equivalent to permeability in a water system). To determine storativity, leakage or porosity, a test involving multiple wells is required. To evaluate these parameters, a single well is pumped and responses over time are analyzed in observation wells. Figure 3 illustrates a typical pump test cross section. Without detailing individual methods of aquifer test analysis, the most important factor is to design a specific test and analysis to meet governing assumptions while providing the necessary data. Aquifer testing requirements should be incorporated when the wells are located and designed during Phase III. Methods of recording aquifer test data range from electric well tapes and stop watches to strip chart records and pressure transducer signals plotted on a field computer. The project engineer is again reminded that contaminated ground water removed from the system may need to be properly contained and disposed.

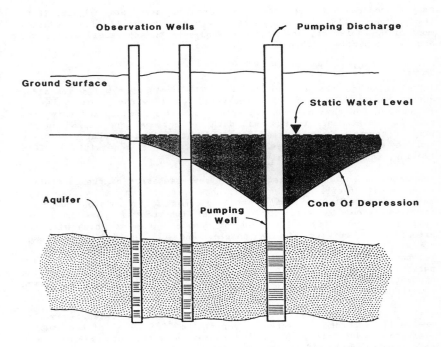

FIGURE 3 TYPICAL PUMPING AQUIFER TEST SCHEMATIC

Many times the costs of collecting enough data to predict contaminant movement are prohibitive. In these cases, computer simulation efforts can provide the required information. The necessary modeling data input may require more sophisticated testing, such as the measurement of porosity, dispersion, sorption, and retardation. To quantify these variables, injection or pumping of a "tracer" solution between two test wells must be conducted. Changes in the tracer solution concentration are measured and used to calculate these parameters. This type of testing is usually relegated to large detailed studies.

Hydraulic data can also be obtained from the open borehole during drilling. Packer testing involves use of a down-hole tool which inflates one or two packers to seal off a specified testing zone. Gas or water pressure is applied to the zone and the response is recorded and analyzed. Packer testing allows several zones to be tested in a single borehole, but requires an open-hole and is very time consuming. Porosity and hydraulic conductivity information may also be obtained from laboratory analysis of the cores and down-hole geophysical logging results.

Many procedures exist to determine hydraulic parameters. Again, it is the project engineer's responsibility to select the field procedures, monitoring equipment, and method of analysis to accurately define the aquifer properties within the project scope.

CHEMICAL QUALITY - PHASE V

In addition to the aquifer testing, the monitoring wells installed during Phase III provide the access which allows the ground water to be sampled for chemical constituents. Sampling frequencies vary from one-time assessments to seasonal data collections, and depend on the project scope. The analysis of ground water chemical characteristics can be divided into three categories: Sample acquisition, field testing, and laboratory analysis.

The acquisition of a sample which is representative of the monitoring zone is imperative. Several methods are currently available to extract ground water samples ranging from Teflon piston pumps to stainless steel bailers. The two most important factors in sample collection are: (1) Utilization of sample equipment which will not affect the chemistry of the sample, and (2) Obtaining a "fresh" formation water sample. Equipment decontamination techniques such as solvent rinsing are usually necessary to eliminate cross contamination. Severely contaminated sites may require dedicated sampling apparatus at each well. Decontamination materials may be deemed hazardous and must be handled appropriately.

Field testing generally is limited to the analysis of parameters which are easily recorded and serve to verify the "freshness" of the sample. Electric conductivity, pH, and temperature represent typical field measurements which aid in the documentation of "fresh" formation water versus stagnant well bore water. Field instrumentation can range from these simple hand-held meters to on-site portable laboratories utilizing GC/MS equipment to provide real time analysis. Procedures for conducting field measurements should be specifically outlined in the QAPP, including instrument calibration, sample procurement, and decontamination.

Drinking water standards and other health effect limits often dictate the quantification of very low contaminant concentrations (in the parts per billion range). In order to assess contaminants at these levels, laboratory analysis is required. Once a representative sample is obtained, specific bottling and field preservation procedures must be followed as documented in the QAPP. Different groups of parameters require different bottle types, filtering, preservatives, and temperature adjustments. Table 2 presents the preservation methods required by the U.S. EPA (1984). Packing and transportation of samples must correspond to required holding time limits and laboratory extraction deadlines. Once the samples arrive at the laboratory, the quality assurance aspects shift to the internal QA/QC program at the laboratory. Prior to the selection of an analytical facility, its internal program should be researched for QA/QC compliance levels which are required for the project.

TABLE 2
FIELD PRESERVATION REQUIREMENTS FOR
GROUND WATER SAMPLING

PARAMETER	SAMPLE CONTAINER	PRESERVATION	HOLDING TIME
Metals	One, 500 ml polyethylene	Filtered (0.45 u) and preserved to a pH <2 with HNO_3	6 months
Cyanide	One, 500 ml polyethylene	Preserved to a pH >12 with NaOH	14 days
Oil and Grease	One, 1000 ml wide mouth glass with aluminum foil under cap	Preserved to a pH <2 with HCL	28 days
General Chemistry	One, 1000 ml plastic	Unpreserved	Varies*
Acid, Base/Neutral Extractables, Pesticides, and PCBs	One, 1 gal. amber glass — no headspace with teflon lined caps	Unpreserved	2 days until extraction
Volatiles	Two, 40 ml glass vials — no headspace	Unpreserved	7 days

Note: All samples should be stored at $4^{\circ}C$.

*48 hrs. to 28 days, dependent on parameter measured.

Laboratory analysis of specific zones of the soil/rock core samples for residual contaminants also may add to the data based required to characterize the ground water resource.

CONCLUSION

The phased approach presented to systematically solve a ground water contamination problem can be utilized for very large or smaller projects by increasing or decreasing the activities in each phase. This targeted approach provides a progressive sequence to focus on specific problem areas. An overview of several data collection techniques has been presented to conduct each phase. Although many data collection procedures are currently available, the project engineer must design any ground water investigation to be responsive to site specific conditions and the needs of the client. A successful program ultimately hinges on the integrity and quality of the information obtained during the investigation.

REFERENCES

Acker, W.L., 1974, Basic Procedures for Soil Sampling and Core Drilling, Acker Drill Company Incorporated, Scranton, PA.

Driscoll, F.G., 1986, Groundwater and Wells, Johnson Division, St. Paul, MN.

Griffiths, D.M. and P.F. King, 1965, Applied Geophysics for Engineers and Geologists, Department of Geology, University of Birmingham; Pergamon Press.

Kruseman, G.P. and N.A. DeRidder, 1983, Analysis and Evaluation of Pumping Test Data, International Institute for land Reclamation and Improvement, Wageningen, The Netherlands.

Lillesand, T.M. and P.W. Kiefer, 1979, Remote Sensing and Image Interpretation, John Wiley and Sons; New York, New York.

U.S. EPA, 1980, Interim Guidelines and Specifications for Preparing Quality Assurance Project Plans, PB83-170514, QAMS-005/80.

U.S. EPA, 1984, Code of Federal Regulations, Vol. 49, No. 209, 40 CFR 136, October 26, 1984, pp. 43260.

Technical Issues of Ground Water Data

Olin C. Braids, Ph.D., Associate*
Gregory K. Shkuda, Ph.D., Senior Scientist*
Gisella M. Spreizer, Scientist*

ABSTRACT

Three fundamental issues to be considered in the design of ground
water monitoring programs are 1) choice of parameters for
characterizing contaminant sources, 2) determining the areal and
vertical placement of monitoring points, and 3) the use of data
collection and analysis procedures to distinguish temporal trends in
water quality. Parameters chosen for ground water analysis should be
characteristic of the contamiant source, amenable to the selected sampling method, and available as part of common commercial laboratory
services. The siting of the borings, and the determination of well
screen elevations should be based on a real-time water quality monitoring program operated in conjunction with the drilling. Drilling
techniques which allow samples to be collected as drilling progresses
should be given preferential consideration. These considerations will
allow individual geologic strata to be sampled and thus allow wells to
be installed optimally. Monitoring programs should have overall data
analysis goals in mind, and the data collected must be adequate to
evaluate the goals of the program.

- - - -

Proper planning of ground-water monitoring investigations is
crucial to the collection of meaningful data which will satisfy program
goals. Two fundamental issues should be addressed and incorporated
into all programs. The first is the selection of monitoring parameters
which will efficiently characterize the water quality. The second is
selection of appropriate statistical tests so that the maximum amount
of information can be extracted from the monitoring data.

The most direct way of choosing water quality parameters is
simply to adopt those that are dictated by regulations. For example,
the RCRA regulations stipulate three sets of parameters that should be
chosen for ground-water monitoring. One is the list of primary EPA
drinking water constituents. This list includes both inorganic and
organic pesticide compounds. Another set of parameters is provided for
the establishment of general ground-water quality (see May 19, 1980
Federal Register, p. 33257). There are five inorganic and one organic
constituents in this set. Finally, there are four parameters, pH,
specific conductance, total organic carbon and total organic halogen
that are specified as indicators of ground-water contamination. The
combination of these water quality constituents would provide a

*Geraghty & Miller, Inc., 125 East Bethpage Road, Plainview, NY 11803

51

reasonably good estimate of ground-water quality in the area of the
facility.   It also should provide an indication of the facility's
influence on ground-water quality.   However, the use of these
parameters does not provide enough information to identify the specific
kinds of compounds that may be emanating from a waste holding or
disposal facility.

To characterize the influence of a waste or storage facility on
ground-water quality at least one and preferably more components of the
materials that are being disposed of or stored must be selected as an
analytical parameter.   If the facility is located on a plant site, the
characterization of the materials held in the facility would be less
complex than if the facility were being utilized for disposal by
multiple industries.   Even if a company produces multiple products,
there is still a limit to the numbers of chemical compounds which are
likely to be present either in byproduct streams or in other wastes
stockpiled at the site.   Since inventories and records of materials
manufactured or handled should be available, a chemist should review
that information and decide upon a listing of appropriate chemical
monitoring parameters that are consistent with the operations being
investigated.

The priority pollutant chemical list which is used universally as
a screening tool for analytical purposes was actually developed from
data obtained from industrial effluents.   As a result, many of the
substances in general industrial use, appear on the list.   In choosing
parameters to reflect the materials being investigated, one of the concerns is to choose compounds for which analytical methods exist.   The
use of the priority pollutant analysis is a good first approach to
characterization of ground-water contamination.

When dealing with industrial or hazardous waste products there is
the potential that uncommon compounds may be present in the waste
stream because the waste product(s) may actually consist of process
byproducts.   However, it is not usually necessary to have an absolutely
complete chemical characterization of the system in order to determine
whether contamination exists.   The organic chemical constituents may be
categorized, as is done with the priority pollutants, in terms of an
analytical scheme that separates volatiles from semi-volatiles and non-
volatiles.   They may also be classified according to analytical type
such as base/neutral-extractable, acid-extractable, or by chemical
type, e.g. alcohol, ketone, aldehyde, or acid.   To determine if contaminants are present in the ground-water system, a chemist would choose
from the list of potential contaminants those which would be stable
(not be degraded), the most highly water soluble, the most mobile, and
the most easily analyzed for.   These compounds would then comprise the
suite of contaminant indicators to be used to study the system.

For situations where wastes have been collected from a
multiplicity of industries, the potential for a wide variety of
compounds is greater.   Under these circumstances, the priority
pollutant analytical scheme would be expected to adequately
characterize a majority of the contaminants expected to be present.   It
would be unusual for a combination of wastes to be so exotic as to
contain only compounds that are excluded from the priority pollutant

list. However, in the unlikely event of this occurring, EPA has taken
a further step in terms of characterizing wastes by developing what is
now termed Appendix IX in the RCRA ammendments. Appendix IX is a list
of hazardous materials which number over 300 compounds and which are
specified as analytes of choice when contamination is discovered at a
hazardous waste site. The approach to the analysis for Appendix IX is
similar to that of the priority pollutants. The compounds and elements
are divided into chemical classifications which would be looked for
initially in a generic sense and then followed by a more specific
analysis if their presence is indicated. Appendix IX was developed as
a screening approach for situations where the more conventional
priority pollutant analysis is deemed inadequate to characterize the
pollutants.

       An attempt should be made to provide some internal analytical
quality control when choosing parameters for ground-water monitoring.
Parameters such as total dissolved solids and conductivity should be
included together so that they can act as an internal performance
check. Calcium and magnesium coupled with hardness will also give an
indication of the quality of the analytical work. The inclusion of
major cations and anions even when the contaminants in question may be
heavy metals or organic compounds, will permit the calculating of a
cation/anion balance and thus monitor the quality of the overall
analysis.

       The ability to characterize the influence of a waste management
facility on ground water, goes beyond a simple analysis for components
in that waste facility. The overall regional geochemical conditions
need to be determined and taken into account. This is most important
if any parameters are present in significant concentrations in the
background geochemical environment. There may be special conditions in
the ground-water system involving salinity, pH, or dissolved oxygen
which could complicate the interpretation of data and require that a
more complex analysis of the geochemical results be included than might
be considered at first glance.

       When designing monitoring programs, the overall project
objectives must be considered, and the data collected must be adequate
to be able to satisfy these objectives. These concepts seem basic, but
their implementation can become involved. The adequacy and limitations
of the collected data, as well as the limitations of statistical tests
anticipated to be applied, should be incorporated in the program's
design. The experimental design will ultimately determine what
objectives can be met, or at least investigated. Investigating a
question does not necessarily mean that it can be answered.

       Statistics can be used to help quantify the degree of uncertainty
in conclusions based on the experimental data. There are four major
sources of variation inherent in all data collected from ground-water
monitoring programs. They are spatial, temporal, sampling, and
analytical variabilities. Spatial variability arises from the fact
that there are differences between wells, even if they are constructed
similarly and monitor the same area of the ground-water flow system.
This variability can be due to geology, hydrology, geochemistry. Most
often the variability exhibited cannot be attributed to any one source.

Temporal variability can be either longterm and monotonic (continually increasing or decreasing concentrations), or short term and periodic (frequent fluctuations between high and low concentrations) and dependent on, for example, seasonality or stresses on the aquifer system. Sample collection and subsequent analyses of these samples add additional sources of error to the data.

The relative magnitude of each error source has been assessed by Doctor et al, 1985. The greatest relative variability is attributed to the spatial differences between wells. The second major source of errors is temporal changes in the system under study. The remaining two sources (sampling and analytic) contribute considerably less to the overall data variability. Due to the inherent variability of the sources of error, it follows that the ground-water monitoring program should be designed so that the influences of these sources are minimized and the maximum amount of information may be extracted from the data.

Most statistical concepts can be considered in either qualitative or quantitative terms. Although less effort is involved in the qualitative review of the monitoring data, the quantification of errors present in the results is also important. Concepts used in the analysis of monitoring data are data validation, identification of outliers, comparison of up- and downgradient results and trend analysis.

The results submitted by the laboratory need to undergo some type of quality assurance/quality control (QA/QC) audit to assure the data's precision and also, if possible, the accuracy of the data. The precision or reproduciability of the combined effect of analytical and sampling work can be measured by the use of blind replicate samples. Guidelines have been set to assess the reproducibility of the data. The EPA has undertaken validation studies to assure laboratory QA/QC and the study results which can be applied to precision checks of the monitoring data (see October 26, 1984 Federal Register).

Accuracy determinations are more involved. It would be necessary to prepare samples with known concentrations, which is usually not feasible during a field study. An alternative is to spike the samples in the laboratory with deuterated or other standard compounds to determine the recovery efficiency. In this way matrix effects in the sample can be assessed. Accuracy and/or precision determinations help to assure the data analyst of the validity of using the data in subsequent statistical analyses.

In addition to checks of the data validity, the data should be reviewed for the identification of unusual observations when the results are compared with previous and also other current sampling results. Potential outliers may be visualized by graphing the data. The graph can have time or sampling events as the baseline (independent variable) and observed concentration as the other (dependent) variable. Measurements should be in line with previous analyses. A graph can also be set up to compare the detected concentrations in up- and downgradient wells. Wells located near to each other (screened in the same portion of the aquifer) should have similar concentrations unless

the differences can be explained by other causes. The removal of
unusual observations should be done with caution. All decisions to
reject data should be made with mathematical outlier tests with
consideration of the actual monitoring and sampling program.

Comparisons of "background" results with downgradient results are
made to determine if the facility or site being studied has adversely
impacted the ground-water environment. There is much controversy
concerning the types of statistical tests used to test for these
differences. Currently forms of the Student's t-test are applied but a
form of the F-test and a nonparametric test have been proposed in the
August 20, 1986 Federal Register. The selection of an appropriate test
depends on the distribution (probability) of the data; i.e. does it
follow the normal Gaussian (bell-shaped curve) distribution?

Trend analysis techniques can be applied to distinguish temporal
trends in water quality data. This aids in the evaluation of whether
the ground-water quality at a specific well is improving or degrading.
The first and most important step in any trend analysis is preparing
graphical representations of the data. By doing so a qualitative
evaluation of the data's behavior can be made. These graphs can then
serve as guides to the applicability of various statistical tests and
the appropriateness of the models used to evaluate the data. As with
the comparison of up- and downgradient results, the distribution of the
data must be taken into account in the model selection.

The evaluation of ground-water monitoring data involves keeping
the overall data history in mind. Current results must be compared to
past. Questions, such as the following always need to be answered:
Does graphing the data give an indication of its behavior; is the
current data set in line with expectations; do the detected
concentrations follow any pattern; can unusually high or low
concentrations be attributed to natural causes, floods or draughts; is
there a cyclic pattern parallelling seasonal changes being exhibited by
the data?

The issues presented (selection of monitoring parameters and data
analysis) are two major components in the design of an effective
monitoring program. The effectiveness of the program is directly
related to having an adequate understanding of the characteristics of
the site and also the realization of the limitations and deficiencies
of the available information. The interpretation of program results
must keep in mind the overall data and site histories.

56   CONTAMINATED GROUND WATER

Appendix A

References:

Doctor, L.G., R.O. Gilbert, R.R. Kinnison, Statistical Comparisons of
    Ground-Water Monitoring Data, Milestone 2. Ranges of Variation
    in Ground-Water at Hazardous Waste Sites, Draft, Feb. 25, 1985
    (Pacific Northwest Laboratories, Richland, WA).

Federal Register, "Guidelines Establishing Test Procedures for the
    Analysis of Pollutants Under the Clean Water Act; Final Rule and
    Interim Final Rule and Proposed Rule," 40 CFR 136, October 26,
    1984, pp. 43234-43442.

Federal Register, "Hazardous Waste Land Disposal Facilities;
    Statistical Procedures for Detecting Ground-Water Contamination,"
    40 CFR 264 and 265, August 20, 1986, pp. 29812-29814.

Federal Register, "Hazardous Waste Management System: General," 40 CFR
    260, May 19, 1980, p. 33257.

A CALL FOR NEW DIRECTIONS IN DRILLING AND SAMPLING MONITORING WELLS

Joseph F. Keely* and Kwasi Boateng**

Introduction

There is no ideal monitoring well installation method for all purposes, so one should consider specific conditions at a site before deciding which drilling and development methods to use. The most widely used drilling methods include air and mud rotary methods, the cable tool or percussion method, and augering. Common development techniques include air-lift, surging and bailing, and overpumping. Specialized techniques for installation of monitoring wells at hazardous waste sites have begun to evolve from these conventional installation methods.

Prior to sampling a monitoring well, it is important to purge it of stagnant water in casing storage, so that the well contains only water freshly withdrawn from the aquifer. In wells having significant volumes of water in casing storage above the wellscreen, special measures must be taken to ensure removal of that water (regardless of the number of borehole volumes purged, as will be discussed below). The occasional need to remove sediment that may have accumulated between sampling events can also be satisfied by vigorous purging. The amount of sediment that may have accumulated is directly related to the effectiveness of the filter pack. There is thus a relationship between the quality of the monitoring well installation and the degree of purging needed.

The literature contains sufficient documentation of the operational principles involved in common installation and sampling techniques (Driscoll, 1986; U.S. EPA, 1975; Scalf and others, 1981; Gibb and others, 1981; Schuller and others, 1981; Schmidt, 1982; Keely, 1982; Keith and others, 1983; Fetter, 1982; Barcelona and others, 1984; Barcelona and others, 1985). This is not a review of that material, but is instead a discussion of some of the unspoken limitations of common techniques.

*Assistant Professor, Oregon Graduate Center, Department of Environmental Science & Engineering, 19600 N.W. Von Neumann Dr., Beaverton, OR 97006.

**Project Manager, Roy F. Weston Inc., Geosciences Department, One Weston Way, West Chester, PA 19380.

Mud Rotary Drilling

Mud rotary drilling (Figure 1) has been extensively employed in the drilling of wells for resource development; e.g., water supply wells, oil and gas wells, and geothermal wells. One of the reasons for the popularity of mud rotary drilling is that it is rapid; in excess of one hundred feet of borehole advancement per day is common. Another reason is that many borehole geophysical logs (e.g., resistivity) must be run in an uncased borehole, and mud rotary drilling facilitates this. The use of mud rotary drilling for the installation of monitoring wells at hazardous waste sites has been generally limited to drilling in off-site areas or through formations known to be free of contamination. These limitations are the result of concerns for potential cross-contamination of strata exposed to the mud circulating in the borehole during drilling.

Even these limited uses of mud rotary drilling at hazardous waste sites are fading quickly, because of uncertainties about the ability of well development efforts to remove drilling mud residues. Mud remaining in the formation after drilling may lower the permeability locally, causing certain strata to yield lesser amounts of water than they should. Mud residues may also serve to alter the ground-water chemistry by binding metals, sorbing organics, supporting excessive biological growth, and altering the cation exchange capacity, pH, and chemical oxidation demand of native fluids.

Air Rotary Drilling

This drilling method is primarily limited to semi-consolidated and fully consolidated formations and is similar to mud rotary drilling, except that compressed air is used to cool the drill bit and to carry cuttings to the surface. This air is injected through a swivel hose plumbed into the top of the drill pipe. Small amounts of water or foam are sometimes admixed with the compressed air to reduce dust emissions and to cool the swivel. In completely unconsolidated formations conventional water-based drilling fluids must be used (Johnson Division, UOP, Inc., 1975). To minimize problems with borehole caving in unconsolidated formations, a temporary casing is driven as the borehole is advanced (Figure 2).

Some concerns are evident when using the air-rotary method. An important one relates to the use of air as the drilling fluid. Unless the air is filtered, oils and other contaminants will be introduced from the compressor. If substantially contaminated strata are encountered, the use of high pressure air may pose a significant hazard for the drilling crew due to the potentially rapid transport of dangerous vapors up the borehole during drilling. Another concern relates to the possible use of water under pressure when drilling in unconsolidated formations (the alternative is mud, which is less desirable, as discussed above), because this may cause extensive invasion into the near-borehole environment. Invasion by non-native fluids can result in local pH and pE changes, and associated chemical changes. The invasion of fluids into the formation can result in the

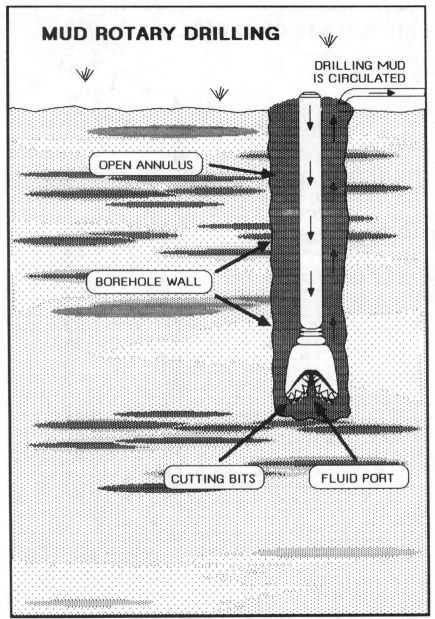

Fig. 1 Cut-Away Sketch of the Mud Rotary Drilling Method

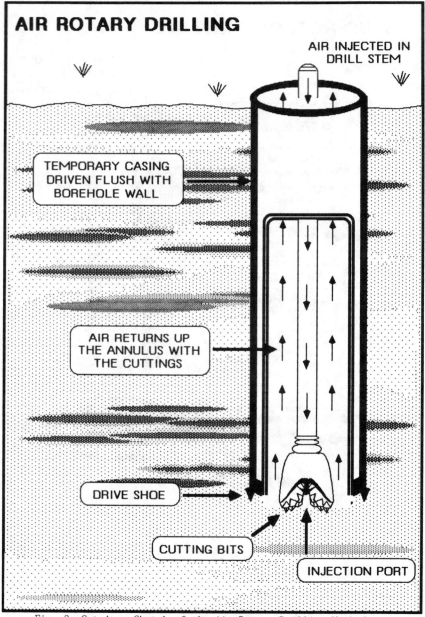

Fig. 2 Cut-Away Sketch of the Air Rotary Drilling Method

need for extensive well development and purging efforts, to ensure that all invading fluids have been removed so that samples taken will reflect the quality of water in the formation prior to drilling. Finally, if foam (usually a surfactant) is used as a drilling fluid additive there will be questions concerning the chemical characteristics and reactivity of the foam.

Cable Tool Drilling

The cable tool method (Figure 3) has been extensively employed for the installation of water supply wells, but has been used only infrequently at hazardous waste sites. Its use may be advantageous in hazardous waste situations, especially where hydrogeologic conditions are variable or not well known. A major advantage is that it allows relatively precise and accurate water sampling during drilling, due to the minimal disturbance of the formation. The data obtained from such samplings helps to determine in which interval to place the wellscreen (Boateng and others, 1984).

In cable tool drilling, the borehole is advanced by lifting and dropping a heavy string of drilling tools that are suspended on a steel cable and terminate in a chisel shaped bit. The impact of the bit breaks up the formation and the resulting material must be removed from the borehole. Cable-tool drilling employs the use of a temporarily emplaced casing that primarily serves to keep the borehole from collapsing during drilling, but that also minimizes possible cross-contamination between strata. In consolidated formations it is not necessary to drive a temporary steel casing to prevent collapse of the borehole, but this may still be useful in contaminant investigations as a means of isolating strata.

The temporary casing is equipped with a sharp drive shoe attached to the lower end, which aids the advancement of the casing by carving out a slightly larger diameter borehole than made by the action of the bit alone. Water is sometimes added (this is usually necessary above the saturated zone) and the loosened material is mixed by the up-and-down action of the drilling tools, to form a slurry for ease of removal by a bailer or a sand pump. The amount of water added is usually small and it is not under pressure, in contrast to water used in the air rotary method. In either case, however, care must be taken to ensure that the water that is used is itself free of contamination.

The process of drilling, driving the casing, and bailing is repeated until the temporary casing is at the required depth. The permanent well casing and screen assembly is then lowered inside the temporary casing to the required depth, and the process of filter packing and grouting is carried on while gradually withdrawing the steel casing. The latter process takes some experience and skill because the filter pack may bridge and result in pulling the permanent well casing out with the temporary casing, and because the filter pack and grout tend to slump in place. The techniques found to be most useful in avoiding such difficulties are:

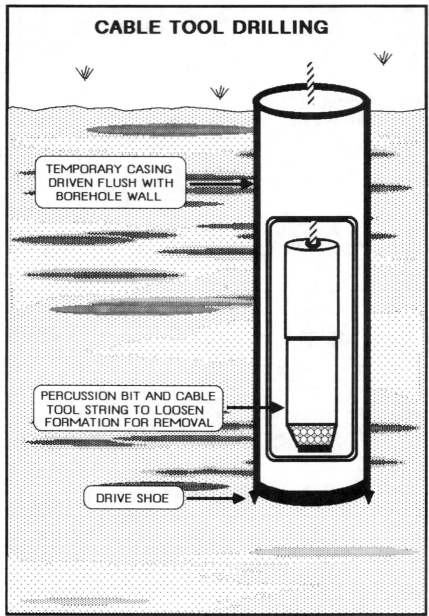

Fig. 3 Cut-Away Sketch of the Cable Tool Drilling Method.

(1) use of a temporary casing of sufficiently large diameter so that an annulus of three inches or more surrounds the permanent well when it is placed in the temporary casing,

(2) use of fine, uniform gravel for packing material to minimize the potential for bridging (often a problem with pea-gravel) and to reduce the amount of well development and purging necessary, by more effective filtering of the aquifer fines,

(3) allowing for considerable slumping of emplaced materials by adding at least thirty percent more than calculations of the borehole volume indicate are minimally necessary to fill the annular space between the temporary casing and the wellscreen,

(4) emplacing the filter pack or grout, and withdrawing the temporary casing, in two- to five-foot stages, and

(5) use of a tri-cone bit or under-reamer during drilling, for those clay strata that are dry or nearly dry when encountered, so that the temporary casing will not be trapped in expanding clays as they become wetted by further work in the borehole.

The annular space between the formation and the temporary casing is ordinarily negligible with cabletool drilling, unlike augering. The amounts of water added to the borehole (when drilling through the unsaturated zone) are usually small and tend to remain inside or close to the temporary casing. Hence, there is little likelihood of the potential cross-contamination that might occur by moving contaminated fluids or solid material from one stratum to another. Driving and removing the temporary casing does not result in significant disturbance of the sediments encountered because of the smoothness of the casing and the slowness of its advancement. This can be a major advantage in fluvial and glacial sediments, which typically have interstratified silt and clay lenses that tend to smear with augering. Another useful advantage to the traditional techniques of cabletool drilling is that a sufficiently large diameter borehole may be drilled so as to enable multiple-well completion within a single temporary casing. This is an extremely complicated task to carry out successfully, however, and is advisable only when drilling several adjacent boreholes would not be feasible economically (e.g., great depth) or technically (e.g., limited space).

The major disadvantages of the cabletool method are the time and costs involved. It takes an average of one field day to withdraw an 8-inch diameter steel casing from a depth of about 100 feet in an unconsolidated formation if hydraulic jacks are not used; where expanding clays are present, we have observed progress as poor as a few tens of feet per field day. The nominal cost typically exceeds $20 per foot of depth drilled; however, this cost must be viewed in balance with the lowered costs gained by the need for less development of the well constructed, which results from the ability to emplace a large and effective filter pack.

Hollow-Stem Augering

Hollow-stem augering (Figure 4) is fast and relatively inexpensive. Several hundred feet of borehole advancement per day in unconsolidated sediments is possible. The cost per foot of borehole is about $10-$15. These factors alone make it competitive with rotary methods. For ground-water contamination investigations it is preferred over rotary methods, because:

(1)   no drilling fluid need be used,

(2)   it is believed that each stratum encountered may be prevented from contacting fluids or solids from other strata (see criticisms below),

(3)   solid samples are readily retrieved by split-spoon samplers during the course of drilling, and

(4)   one can remove the center plug when the target depth has been reached and immediately begin construction of the monitoring well inside the hollow auger flights.

In the augering method, the hole is advanced by rotating and pressing the auger into the soil. As the auger is pressed into the soil, cuttings are rotated upwards on the auger flights. This poses a potential cross-contamination problem since contaminated material from a lower stratum may be brought into contact with an uncontaminated overlying stratum. The augering action also causes interstratified clays and silts to smear into open sand and gravel strata, possibly changing local permeabilities and affecting the proportion of flow delivered from each stratum to the monitoring well.

To ensure that the auger does not become bound to subsurface materials, binding-prevention techniques are employed. These generally consist of rotating the auger flights in place, or sequentially raising and lowering the flights a few feet while they are rotated; ordinarily this is done every few feet of borehole advancement. Such actions aggravate the smearing of clays and silts into other strata, because the vigorous action of the rotating flights pushes loosened material against the borehole wall. Moreover, binding-prevention techniques enlarge the borehole beyond the nominal diameter of the auger flights. Thus, potentially contaminated solids or fluids from an overlying stratum may be brought into contact with a lower stratum by falling down the annular space.

The major limitations of this drilling method are that it cannot be used to drill hard rock formations and is generally incapable of being used to drill much deeper than a hundred feet. Augering is also slow in coarse materials such as cobbles and boulders. One technique found useful to minimize the difficulties in augering through such sediments is the use of a small diameter tri-cone bit (run through the

HOLLOW–STEM AUGERING

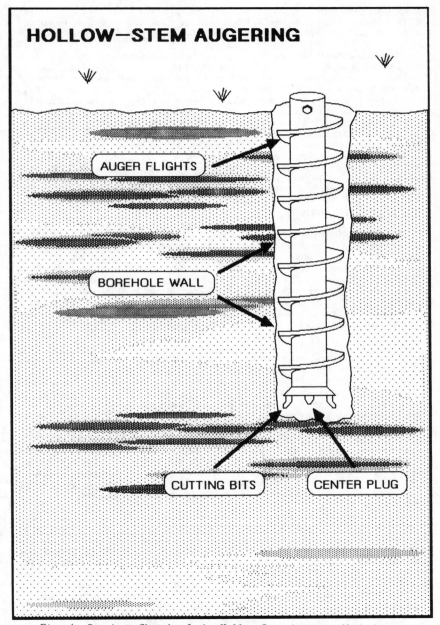

AUGER FLIGHTS

BOREHOLE WALL

CUTTING BITS   CENTER PLUG

Fig. 4 Cut-Away Sketch of the Hollow-Stem Augering Method.

center of the auger after removing the plug) to shatter the large cobbles encountered. No drilling fluid is needed since the amount of drilling with a tri-cone bit is minimal for this purpose.

Obtaining representative water samples from discrete strata during augering is possible, though special modifications are needed; such as the Keck Screened Method that incorporates a screened section into the shaft that the auger flights are welded to (Scalf and others, 1981). Hollow-stem augers are particularly susceptible to 'sandblows'(or 'heaving'), wherein the removal of the central plug is accompanied by an immediate rise of loose sediments inside the lower auger flights. This can be quite severe (i.e., several tens of feet of sediments) and complicates the retrieval of representative solid samples. The occurrence of sandblows can be verified by measuring the length of pipe that is lowered with split-spoon samplers, Shelby tubes, and core barrels, so as to be certain that the total depth of the borehole has been reached before the solid sampling procedure begins.

A Hybrid Drilling Method

The foregoing discussions indicate that none of the conventional drilling methods is without its technical or economic disadvantages. Clearly, the rapid advancement of the borehole by augering and mud rotary drilling, and their relatively low costs, are desirable features of these techniques. The complete isolation of strata using temporary casing, the superior capacity for filter pack emplacement, and the ability to sample the aquifer at specific intervals are key advantages of the air rotary and cabletool methods. It is only reasonable to want all these advantages in a single technique.

Figure 5 illustrates one possibility that has been field tested by the author and found to be satisfactory. The use of temporary casing is borrowed from the air rotary and cabletool methods, while borehole advancement is accomplished by either an auger or a mud rotary drill that is run through the center of the temporary casing. Augering is preferred because no foreign fluids need be introduced. This is not essential, however, because clean water may be used with mud rotary drills. A heavy drilling mud is not required to stabilize the borehole walls, because this is taken care of by the temporary casing. In either case, the borehole is advanced a foot or two at a time, followed by driving the temporary casing to the new bottom-hole depth (driving the casing is accomplished by a casing hammer or a drop-weight attached to a rope run around the cat-head of conventional drilling equipment; per the usual techniques for driving solid samplers into the formation).

There are two major reasons for selecting hollow-stem augering as the preferred method of borehole advancement. First, if a hollow-stem auger is used, it may facilitate the rapid acquisition of subsurface sampling with common devices like split-spoon samplers and Shelby tubes. It is usually easier to withdraw the central rod and bottom plug from a hollow-stem auger than it is to withdraw the entire drill stem on a rotary drill rig. Secondly, the use of any drilling fluid is

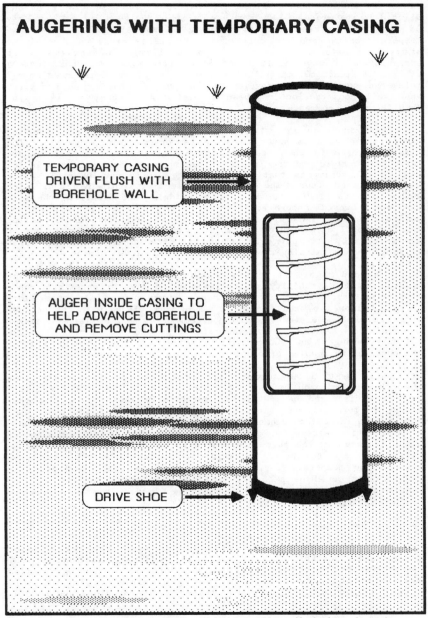

AUGERING WITH TEMPORARY CASING

TEMPORARY CASING
DRIVEN FLUSH WITH
BOREHOLE WALL

AUGER INSIDE CASING TO
HELP ADVANCE BOREHOLE
AND REMOVE CUTTINGS

DRIVE SHOE

Fig. 5 Cut–Away Sketch of a Hybrid Drilling Method –– Augering
Within a Temporarily Driven Casing.

68   CONTAMINATED GROUND WATER

cause for some concern, since the native fluids are displaced and diluted in the process. When rotary drilling, quite a lot of fluid may be required to keep cuttings moving up the annulus between the temporary casing and the drill rod stem. This means that obtaining samples of native fluids during the drilling process (e.g., for screening-level determinations of which strata may be contaminated) will be difficult to impossible. It also gives rise to the need to re-think the extent of well development and purging that may be necessary when well construction has been completed with rotary drilling and sampling is to commence.

There are some potential difficulties to consider when advancing the borehole with an auger inside a temporary casing, however. The first of these is the potential for bridging of loosened sediments between the auger and the inside of the temporary casing; in alluvial and glacial sediments that typically have widely varying grain sizes this may indeed occur. The experience gained with this hybrid drilling method thus far indicates that, while bridging has occurred, the rigs used were powerful enough to grind away and free the auger without too much difficulty. Helpful techniques in this regard included reversing the rotational direction of the auger, adding or releasing pressure on the flights by attempting to sequentially raise and lower the auger flights, and brutalizing the assemblage with a sledge hammer.

A second major difficulty that can occur with this technique relates to the inability of hollow-stem augers to penetrate zones containing large cobbles or boulders, since there is little side-slippage allowed by the temporary casing. As found with the hollow-stem method, the easiest way to overcome blockages by large diameter materials has been to remove the center rod and bottom plug from the hollow-stem auger, and run a small tri-cone roller bit down to shatter the obstructing cobbles or boulders.

This same temporary use of a tri-cone bit can be used to alleviate the other major problem so far encountered with the hybrid drilling technique discussed here; that of potentially locking the temporary casing in formerly undersaturated, swelling clays. A small tri-cone bit (two to three inch diameter) is allowed to 'bang around' just ahead of the entire assemblage, gouging out a couple of extra inches of annular space for the temporary casing to be advanced through. Large bits may gouge too much, potentially creating undesirable pathways between the temporary casing and the borehole wall. Obviously, an under-reamer would be an ideal tool for this work, but its use may require that the entire auger assembly be removed from the temporary casing beforehand.

Sampling Techniques

Concerns about the effects that sampling devices may have on contaminant concentrations as a result of their mechanical operational characteristics have been investigated previously, including detailed studies of the potential for degassing or air-stripping volatile organic chemicals (Barcelona and others, 1984; Barcelona and others,

1985). The concern over losses from air-stripping arose quite
naturally, as those who have been in the presence of a major production
well while it is producing water laden with volatile organics can
attest; enough of the contaminants are volatilized to constitute an
odor and health exposure problem.

There are, however, substantial differences between the masses of
contaminants volatilized by high discharge rate production wells per
unit time and the masses that are volatilized to purge monitoring wells
prior to sampling. In the extensive comparison of purging and sampling
devices conducted by Barcelona and others (1984), bladder pumps ranked
above average in overall sampling performance, conventional bailers
ranked average, and mechanically-driven positive-displacement type
pumps (such as the Johnson-Keck SP81^{tm} stainless-steel pump) ranked
average to below average. More specifically, that study rated bladder
pumps superior to bailers, and bailers superior to suction pumps, in
controlling the loss of volatile organic chemicals. Unfortunately, the
performance of mechanically-driven positive-displacement pumps, such as
the SP81^{tm}, in controlling the loss of volatile organics was not
reported there. However, dissolved oxygen and methane losses were
examined and found to be significant; one might infer that volatile
organic chemical losses would also be significant. For the other
devices, increasing biases with increasing concentrations of volatiles
were noted.

During the course of investigations at the Chem-Dyne Superfund
site, the first author conducted an ad hoc experiment to determine the
significance of the bias introduced by using the Johnson-Keck SP81^{tm} to
collect volatile organic chemical samples. That pump was already in
use at the site for purging the monitoring wells and for collection of
samples to be analyzed for metals, inorganic species, pesticides, and
non-volatile organic chemicals on EPA's Priority Pollutant List. In
the usual sampling, when purging was complete, the pump was allowed to
continue discharging for collection of those samples; then the pump was
quickly removed and a stainless-steel bailer lowered to collect the
sample for volatile organic analyses.

On one occasion, however, samples were taken for volatile organic
analyses from a few wells using the discharge of the SP81^{tm} pump;
followed immediately by sampling in duplicate with a bailer. The
results of this ad hoc field experiment are shown in Figure 6. For the
well that exhibited high levels of volatiles (well MW10), it is hard to
distinguish the samples obtained with the SP81^{tm} pump from those
obtained with the bailer. For those wells that had very low (well MW-
12) to moderate (well MW-13) levels of volatiles, the SP81^{tm} samples'
total volatile concentrations are greater than those reported for the
original bailed samples. As a whole, these data do not follow the
trends reported by Barcelona and others (1984). There is only mild
indication of bias with the SP81^{tm} and it is not persistent; it does
not increase with increasing concentrations. These results may indeed
provide evidence that adequate purging can remove the greatest source
of imprecision in reported data.

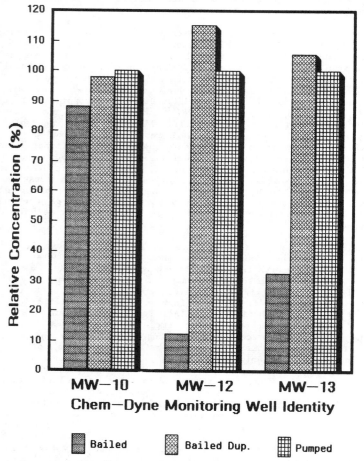

Fig. 6   Overall Relative Comparison of Bailed and Pumped VOC
Samples Obtained from Chem—Dyne Monitoring Wells.

Having worked in or dealt with analytical laboratories for many years, it is recognized by the authors that the data presented here is far from definitive. A more thorough characterization of the mechanisms producing volatile losses in mechanically-driven positive displacement (and other) devices is needed. In addition to controlled laboratory experiments, much larger surveys of on-site experiments are needed.

Such research may have consequences for the many contaminant investigations that continue to incorporate data from small private and public production wells in mapping the extent of plumes. Some method of properly normalizing such data to other data collected from monitoring wells with recommended devices, such as bladder pumps or bailers, must be found.

CONCLUSIONS

Selection of proper monitoring well drilling and sampling techniques cannot be over emphasized. Every technique has its advantages and disadvantages; therefore, hydrogeologic conditions and project objectives should be carefully evaluated before deciding which field techniques to use.

The hollow-stem auger drilling technique has an inherent limitation of disturbing large volumes of subsurface materials around the borehole, thereby possibly affecting local permeabilities and creating annular channels for the movement of contaminants into previously clean strata. At hazardous waste sites where hydrogeologic conditions are not well defined, or where a known or suspected contaminated aquifer is perched above an important aquifer, the hollow-stem augering technique may not be advisable. Hollow-stem augers often does not allow for emplacement of an adequate filter pack. This may result in excessive well development and purging requirements.

Air rotary and cabletool drilling use a temporary casing that shields one strata from another, and allows the emplacement of an adequate filter pack. Air rotary drilling may result in the emission of noxious vapors from the borehole when contaminated strata are encountered, however; and may result in undesirable invasions of the near-borehole environment when water is used. The cabletool drilling method is often slow and expensive. The hybrid technique described here may offer the advantages of both of the preceding methods without their disadvantages.

The recommendations given to date regarding sampling devices may not have adequately considered the performance of mechanically-driven positive-displacement devices. A limited field test indicates that more work should be done in this area. Such work may have implications for the treatment of data from private wells, that are often included in contaminant plume definition efforts.

ACKNOWLEDGEMENTS

We are grateful to Dr. Michael Barcelona of the Illinois Department of Energy and Natural Resources Water Survey Division, Ms. Emma Hampton of U.S. EPA's Hazardous Waste Engineering Research Laboratory (Cincinnati, Ohio) and Mr. Steven Young of Roy F. Weston, Inc. (West Chester, Pennsylvania) for helpful comments. Thanks go to Ms. Kathy Clinton for assistance in creating the illustrations of drilling methods. The material presented in this discussion is condensed from two documents that have been accepted for publication in <u>Ground Water</u> (Keely and Boateng, 1987a and 1987b). The author is grateful to the Association of Ground-Water Scientists and Engineers for permission to release this condensed version.

DISCLAIMER

Although this article was written in whole or in part by employees or contractors of the U.S. Environmental Protection Agency, it has not been subjected to Agency review, and therefore does not necessarily reflect the views of the Agency; no official endorsement is inferred. The mention of tradenames does not constitute endorsement of any kind.

REFERENCES

Barcelona, M.J., J.A. Helfrich,E.E. Garske and J.P. Gibb, 1984. A laboratory evaluation of ground water sampling mechanisms. Ground Water Monitoring Review, v.4, no.2, pp. 32-41.

Barcelona, M.J., J.P. Gibb, J.A. Helfrich and E.E. Garske. 1985. Practical guide for ground-water sampling. U.S. EPA Office of Research and Development, publication no. EPA/600/2-85/104. U.S. EPA, R.S. Kerr Environmental Research Laboratory, Ada, OK, 169 pages.

Boateng, K., P.C. Evers and S.M. Testa. 1984. Ground water contamination of two production wells: a case history. Ground Water Monitoring Review, v.4, no.2, pp. 24-31.

CH_2M Hill. 1984. Remedial investigation report: chem-dyne site, hamilton, ohio. Unpublished report on contract no. 68-01-6692 to U.S. EPA Region 5 office, Chicago, Illinois, (numbered by chapter).

Driscoll, F.G. 1986. Ground water and wells, second edition. Johnson UOP, Inc., St. Paul, Minnesota, 440 page.

Gibb, J.P., R.M. Schuller and R.A. Griffin. 1981. Procedures for the Collection of Representative Water Quality Data From Monitoring Wells, Illinois State Watery Survey, Illinois State Geological Survey and U.S. Environmental Protection Agency. Cooperative Ground Water Report No. 7, pp. 5-7 and 26-28.

Keely, J.F. 1982. Chemical time-series sampling. Ground Water
    Monitoring   Review, v.2, no.4, pp. 29-38.

Keely, J.F. and F. Wolf. 1983. Field applications of chemical time-
    series   sampling. Presented to the Third National
    Symposium on Aquifer Resotration and Ground Water Monitoring,
    Columbus, Ohio, May, 1983; and published in Ground Water
    Monitoring Review. v.3, no.4, pp. 26-33.

Keely, J. F. and K. Boateng. 1987a. Monitoring well installation,
    purging, and   sampling techniques - part 1:
    conceptualizations. Ground Water, v.25,   no.3.

Keely , J. F. and K. Boateng. 1987b. Monitoring well installation,
    purging, and   sampling techniques - part 2: case
    histories. Ground Water, v.25, no.4.

Keith , S.J., L.G. Wilson,H.R. Fitch, and D.M. Esposito. 1983. Sources
    of   spatial-temporal variability in ground-water quality
    data and methods of control. Ground Water Monitoring Review,
    v.3, no.2, pp. 21-32.

Miller, G.D. 1982. Uptake and release of lead, chromium, and trace
    level   organics exposed to synthetic well casings.
    Proceedings of the Second National Symposium on Aquifer
    Restoration and Ground Water Monitoring. National Water Wells
    Association, Columbus, Ohio, pp. 236-245.

Scalf, M.F., J.F. McNabb, W.J. Dunlap, R.L. Cosby and J.S. Fryberger,
    1981.   Manual of ground-water sampling pprocedures. U.S.
    Environmental Protection Agency, pp. 43-71. U.S. EPA, R.S
    Kerr Environmental Research Laboratory, Ada, OK, (93 pages).

Schuller, R.M., J.P. Gibb and R.A. Griffin. 1981. Recommmended
    sampling   procedures for monitoring wells. Ground Water
    Monitoring Review, v.1, no.1, pp. 42-46.

Schmidt, K.D. 1977. Water quality variations for pumping wells.
    Ground   Water, n. 15, no.2, pp. 130-137.

Schmidt, K.D. 1982. How representative are water samples collected
    from wells?   Proceedings of the Second National Symposium
    on Aquifer Restoration and Ground Water Monitoring. National
    Water Wells Association, Columbus, Ohio, pp. 117-128.

Wolf, F. and K. Boateng, 1983. Report of the ground-water
    investigation,   Lakewood, Washington. October, 1981-
    February, 1983. U.S. EPA Region 10 Office, Environmental
    Support Division, Seattle, Washington.

Managing Ground Water Data

Edward Kaplan (member ASCE) and A. Meinhold*

Abstract

Decisions concerning ground water protection and public health should be based on all relevant data, measured and analyzed with technically valid techniques. In areas where ground water quality is of concern, a large amount of ground water data may exist at several agencies, in many forms. This data is likely to be of variable quality, with differing spatial and temporal characteristics. A methodology is presented for identifying and integrating available ground water data.

Introduction

Private firms and government agencies must often make decisions concerning capital expenditures, public health, and compliance and other regulatory issues on the basis of whatever ground water data are readily available. Ideally, such decision making should be based on all relevant data, which have been measured and analyzed with technically valid techniques. However, ground water data present problems in both availability and analysis. In any area or region where the quality of ground water is of concern, important data are likely to exist at many agencies and in many forms. Therefore, making such data available to the analyst or decision maker is a difficult task. Analyzing and assessing ground water data are also difficult because of the complexity of ground water systems, the variability of the spatial and temporal characteristics of the data and differences in collection and laboratory procedures. Ground water data are thus "diverse," existing in different forms and places and possessing different characteristics of space, time, and quality. Brookhaven National Laboratory (BNL) has written a guidebook which presents a methodology for identifying and integrating available data in a useful form, and for analyzing these data using statistical, modeling, and graphical techniques (Kaplan et al., 1985). The approach to data collection is summarized here. The methodology is aimed primarily at local agencies with limited available manpower and equipment which are responsible for making evaluations based on a large but possibly imperfect collection of data.

* Biomedical and Environmental Assessment Division, Department of Applied Science, Brookhaven National Laboratory, Upton, New York 11973

74

Ground Water Management

Management of diverse surveillance data has evolved in a fragmented and piecemeal manner. As a result there usually exists extensive data bank resources scattered in various forms at many different agencies. This situation has prevented the full utilization of such data because of reluctance to spend the money and manpower required to keep track of the large volume of information. Decisions concerning land use planning, waste treatment and water supply management have thus fallen short of their intended impact or cost effectiveness because potentially important data have not been used.

Discussion of Commonalities

Data to describe aquifers usually exist because of problems found or anticipated with the quality or quantity of waters drawn from underground sources. Ground water data are not usually collected because of scientific interest alone. Herein lies the problem confronting the planner, regulator, enforcement specialist, scientist, engineer, or other person interested in understanding the nature of an aquifer. To wit, research quality data, if it exists at all, describes standard measurements taken at regular intervals at fixed locations irrespective of whether contamination has been found. But because such information requires a long term commitment of funds, manpower and other resources, it usually exists at only a few locations within a region. In contrast, monitoring wells are often constructed in response to contamination, where information is desired on short notice, or where remediation efforts must be identified and accomplished. There are only a half dozen or so generic categories of contamination-producing activities which have led to the collection of most ground water data that are worth assembling (Table 1, Part A). Irrespective of how an individual pollution event occurred, the most important categories of data include source definition (e.g. the amount and characteristics of the pollutant), subsurface hydrology (e.g. depth to water or piezometric pressure), and some background information on whether a contaminant has been found in the area in the recent past (Table 1, Part B). There will be differing kinds and amounts of such data available in any particular region depending both on the magnitude of contamination sources and the degree to which they fall under the purview of public agencies. Hence, an analyst may assemble data from many agencies, several of whom may have collected the same information, while others may have confined themselves only to the information they are required to collect to meet their own programmatic objectives.

Using commonalities to gather data

Table 2 illustrates one approach to the process of creating an integrated ground water management information system (MIS) using data from several sources. Recognizing commonalities between various categories of pollution problems allows one to take advantage of situations where one agency (or perhaps several agencies) has data describing the same occurrences of contamination. This approach helps one to unravel a situation where administrative responsibilities have shifted among agencies, in which case a complete set of required data

would be available only if one knew, <u>a priori</u>, that one such agency
had information prior to some particular time, with another agency (or
department) collecting the same information later.

Table 1

General Categories of Activities Causing Ground Water
Pollution, and Required Supporting Data

<u>Part A   Activities</u>

     Accidental surface spills

     Land application of fertilizer/pesticides
       o   Agriculture
       o   Horticulture
       o   Home Gardening

     Road runoff
       o   Salt
       o   Gasoline/oil/other petroleum products

     Landfill leachates

     Direct contamination
       o   Cesspools/Septic systems
       o   Underground injection
       o   Faulty well construction/operation

     Excess pumpage and/or decreased recharge

<u>Part B Data Commonalities</u>

Hydrogeology	soil and rock types location of ground waters flow regimes aquifer properties water balances
Present water and land use	location of wells and pumpage demography, zoning locations of permitted discharges locations of potential discharges locations of recharge areas
Future plans/options	population growth projection zoning regulation

Table 2
Suggested Table to Locate Data

Problem Category	Type of Contaminants	Who Collects Data	uses data	Administrative Responsibility
Surface Spills				
Landfills				
Public Water Supplies				
Private Wells				

A good illustration concerns spills of hazardous chemicals. There may be several agencies which require notification of such occurrences: a local or county health office, a state department of transportation, and perhaps one or more federal agencies such as the EPA. Few of these agencies (particularly those with administrative responsibilities) may actually assemble any data, (which may have been collected by some other agency). To complicate the situation even further, those same agencies requiring notification of incidences of contamination (e.g. spills) may not always collect the same types of information. Information such as the number of gallons spilled on a roadway may be forwarded to a department of transportation and may not be available at a county health department. The county health department, on the other hand, is very likely to possess relevant chemical measurements which would not be collected by other agencies.

Another important reason exists for appreciating commonalities between categories of pollution incidents. Namely, the same data requirements and numerical analyses are often required to assess the consequences of contamination from each category. Suppose an assessment is required of organic contamination found at several private residences in some proximity to industrial activities. Insufficient information may exist at one agency (e.g., department of health) for particular substances at the locations of interest, but useful data (e.g., for similar chemicals) may exist at another agency to describe contamination events at nearby locations. It would clearly be of importance to obtain such additional information when a ground water MIS is created, even though the effort may seem extravagant or irrelevant at the time. In fact, even if a table like Table 2 provides no immediate clue as to where such data resides, the very existence of entries showing that other agencies may likely be concerned with the same type of (or similar) information may be useful at some later time. Such a table of commonalities is an invaluable management tool. It forces the ground water analyst to confront basic questions which relate to regional aquifer issues: what types of analyses are contemplated, whether sufficient synoptic (or other) data exist to support these calculations, whether one or more agencies collect the necessary information, and who in these agencies possess the data or are required to approve its distribution.

Development of an Integrated, Diverse Information System

The process of gathering, sorting, and storing diverse data related to ground water begins by describing the attributes of a system for managing such information. This includes defining which information is required, where it is located, how it can be obtained, and how it should be entered into the system. The transformation of the management information system (MIS) to a geographically oriented geographic information system (GIS) is the logical next step in the hierarchy of collecting and using diverse data.

Concepts of Management and Information Systems (MIS)

Data required for assessments of aquifer systems will necessarily derive from numerous diverse data sets. Examples include water quality data from monitoring wells and from private homes; well logs from driller's reports filed with appropriate government agencies; depths to water tables and well yields, compiled by the U.S. Geological Survey and other agencies; land use and census information assembled by local planners; and locations of allowed and accidental sources of discharges to aquifers. Each set of data has a structure which reflects not only the types of data being collected but the needs of the collecting agencies themselves.

Use of information from so many sources can be a formidable undertaking unless the required data are extracted and stored in a systematic way. To do this, one must first determine which information is available from each agency and then which information to extract and use. One must also identify the exact form in which each set of data is available: bound documents, loose reports and memos, tables, graphs, charts, and machine readable (i.e., computer) files. Not all information from all sources will be required. Not all data will be usable because spatial and temporal aspects may differ, because background documentation describing how data were collected may be unavailable, and because the human effort required to extract information from non-machine readable records may be too great.

A system is therefore required to extract, assemble, store, access, and analyze data. For purposes of this report, the term "management information system" (MIS) means a system which may exist as software designed to support interaction among users, a data base, applications software, and the host computer's operating system. Figure 1 provides a conceptual outline of the components of an MIS.

Within the MIS, data sets are stored in memory on the host computer and organized into a logical structure termed the data base (DB) under the control of software called a data base management system (DBMS). Data retrieved from the data base by the DBMS can be processed further by applications software such as simulation models, optimization models, statistical packages, and graphics packages. These manipulations can be controlled by MIS software responding to user instructions. The greater the programming effort that goes into MIS development, the more flexible and straightforward this interface

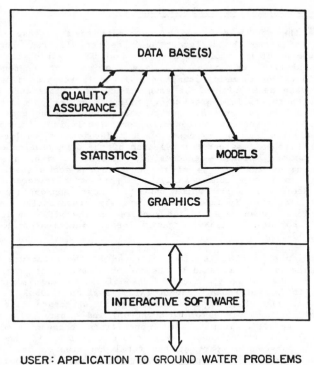

USER: APPLICATION TO GROUND WATER PROBLEMS

Figure 1. Components of a Management Information System (MIS)

can be.   Any specific MIS design will therefore depend on the experience of the user, the complexity of the user's needs, and the inherent capabilities of the DBMS, applications software, and the host operating system.

Geographic location should be the primary concern of the system when the purpose of the MIS is to manage spatial data.   When data organization reflects the spatial assessments that will be made, the MIS is called a geographic information system (GIS) (Marble, Calkins, Peuquet, 1984).

As already noted above, the different types of data, the different agencies responsible for data collection, and the different purposes for which data were collected produce an accumulation of data that are not necessarily comparable.   A methodology is required to relate data so that descriptive and comparative analyses can proceed across jurisdictional, procedural, and natural (physical) boundaries.   The GIS becomes a useful and often necessary step in modeling ground water flow, determining plumes of contamination, and analyzing possible causal relationships between water quality and other spatially distributed data.

Other Information System Considerations

Since the source data which contribute to a diverse MIS are
collected and stored by different agencies for different purposes,
the development of an information system must be flexible with regard
to the quantity of data and its logical arrangement and
manipulation. For example, data may be found in several differently
organized data bases within the same agency, which presents an MIS
designer with the twofold problem of trying to devise an appropriate
logical data structure for his own purposes, while being constrained
by the lack of resources to convert from an existing, incompatible
data organization. The process of MIS development evolves with
continuing feedback, as careful research of the organization of data
at the various source agencies is combined with a better
understanding of the purpose of the new information system. Many
DBMS's are available commercially. Important considerations to
simplify decisions about which DBMS is most appropriate in a
particular situation include: existing data organization, natural
relationships between data, and the purposes of the MIS. An in depth
overview of computer data base organization and function can be found
in Martin, 1977.

Selection of a DBMS is often dictated by the structure of the
incoming data as it is related to the amount of available programming
resources. For example, machine readable data may already be
organized in a particular commercially available DBMS software
package: it then has a particular logical structure (i.e., the
relationship between pieces of information) and a particular schema
(i.e., the specific formats and connections between pieces of
information). It may be less costly to use this DBMS if these data
represent a substantial part of the total anticipated data in the
study, and if considerable programming effort will be required to
convert the data to some other system. In each situation the
programming effort required to convert to a new DBMS must be balanced
against the programming effort necessary to overcome inefficiencies
of existing systems.

Relationships in the assembled data are defined by natural
processes (e.g., physical principles) or by other attributes of the
system under study. These relationships may be hierarchical, some
kind of network, or other structure. An example of a hierarchical
structure is one where data are arranged in terms of a grid: within
each grid square information is stored from those wells located
inside the grid. Each well may in turn be described by information
on its construction and by measurements taken at various times for
each of several chemical constituents in the water sampled from the
well. The data base management system must capture the essence of
this structure to produce the most efficient data storage and
retrieval. An example of a network relationship would be the
association between soils information obtained from a grid located at
some distance from a grid with sampling wells. Some type of causal
relationship may exist between the two grids that would be difficult
to classify as a hierarchy. A DBMS capable of only hierarchical
logical structure would not easily support such a relationship.

The types of applications intended for the data must also be considered in choosing a DBMS. Efficiency may be lost, for example, if the orientation of the study changes and the DBMS cannot be easily structured for a different approach. For some applications such as the statistical association of parameters, natural relationships (e.g., hierarchy or network) may in fact be an impediment to efficient retrieval of information. A simple hierarchical relationship may suffice if the DBMS is designed for inventory data and is used to write summary reports.

Identification of Required Information and Data Sources

Identification of required information is crucial in the development of the MIS and will determine what data are collected, the data base is structured, and the ultimate utility and how applicability of the system.

Analyses and assessments of impacts related to ground water can be linked by common data needs. These commonalities will include the location of pollution sources and impacted wells, depth to water measurements, and measurements of water quality. Other more specific kinds of data will be necessary depending on the application. Table 3 lists some general applications of a ground water MIS, and the data that may be required for each.

Chemical data will make up a large part of the MIS since aquifer assessments are most concerned with water quality over time and space. Such data will be necessary in analyses concerned with trends or changes in water quality, in time series analysis, in analyses and models concerned with pollutant transport, for analyses concerning water quantity and usage, and in relating water quality to surface discharges or land use. Chemical data will also be necessary to evaluate the quality of ground water from the points of view of regulatory compliance and public health. Measurements of water levels will be required in almost all analysis concerning ground water and can be used to infer direction of water flow.

Hydrogeologic information, including the aquifer tapped, aquifer and soil characteristics, and depth of wells, will be necessary for any analysis which involves tracking a pollutant or comparing data from several wells. Data on pumping rates and periods of downtime will help in analyzing the drawdown and zone of influence of large production wells. Well construction data may be used to assess the representativeness of a sample or the impact of construction techniques and materials on chemical data. If ground water models are to be integrated into the MIS, the data and parameters necessary for the models may in part be derived from well construction data.

Information describing land use is required if one of the goals is to relate land use to water quality for help in zoning or regulatory decisions. Some analyses may utilize general land use classifications and changes over time (i.e., a change in land use from agriculture to low density housing). Demographic data such as population, current zoning, and census classifications, may be required for zoning and planning decisions. Other land use and regulatory data are useful in relating a specific contaminant to a discharge (e.g., describing industrial locations, sewage treatment plants, superfund sites, landfills, recharge basins, and underground gasoline tanks). Data on major chemical or oil spills as well as permit information, may also be useful in this regard. Agencies and groups which may maintain the necessary data are described below.

Table 3   MIS Applications and Required Information

Application	Required Data
Statistical applications concerned with water quality: trends, time series, etc.	Water quality, water levels, hydrology (aquifer, depth)
Relate general land use to water quality for zoning/planning	Land use classifications, zoning and population data, water quality, water levels, hydrogeology
Tracking pollutant to source	Water quality, water levels, meteorology, hydrogeology, sources of pollutants
Modeling	Water quality, water levels, soil and meteorological data, hydrogeology

1. Water Quality, Quantity, and Usage

The United States Environmental Protection Agency (USEPA) and United States Geological Survey (USGS) maintain national data bases which contain data on ground and surface water quality. The EPA data base called STORET (an acronym for STOrage and RETrieval) was established by EPA in the early 1960s and is maintained on a computer in Raleigh, N.C. Its original purpose was to provide a system where any person or agency collecting water quality data can store the information in a computerized DBMS. In return, each such data contributor has access to all other data not protected by special access requirements. The system was quickly expanded to allow those who were data users only (i.e., not storing information) to extract unprotected data. STORET has been expanded to provide various sophisticated retrievals (e.g., irregular geographic areas with restrictions based on periods of record and frequency of sampling).

The USGS maintains a data base called WATSTORE at USGS headquarters in Reston, Virginia. The WATSTORE system is akin to STORET, but contains only USGS water data, and includes data related to streamflow, and ground and surface water quality.

USGS district offices also maintain data concerning water quality
and water levels in their areas. The Water Resources Data published
annually by regional USGS offices will contain some of this
information. Other unpublished information or computerized data not
in WATSTORE may be available at the regional USGS office. Not all
machine readable data is transferred to WATSTORE; this is
particularly true for ground water data.

State, county, and local health departments or environmental
protection departments will maintain records (or data bases) of water
quality in supply and monitoring wells within their jurisdictions.
These agencies may also maintain records regarding water levels in
monitoring wells and sometimes produce water table maps of the area.
They may also have data concerning water quality in wells which
monitor spills or continuous discharges, such as landfills, or leaking
underground storage tanks. Data relating to public water quality
supply wells, monitoring wells, and private wells may exist at
different departments within an agency. Information concerning
pumping rates and areas of influence may be maintained by
environmental protection agencies or by a department of public works.
A large body of water quality data resides with water purveyors, who
are required to monitor the water they provide to the public for many
chemicals. Water purveyors also maintain records concerning the
pumping rate at each well and the area served by their supply wells.

2. Hydrogeologic Information

Some hydrogeologic information relating to specific wells can be
found in the "HEADER FILE" of the USGS WATSTORE data base.
Hydrogeologic data may also be available in paper or computerized form
at the local USGS office. USGS technical papers and open file reports
are available which describe the geology and hydrogeology of many
places in the United States. Some hydrogeologic information may be
available at the state, county, or local agencies or with water
surveyors which handle the associated water quality data.

3. Well Construction and Well Logs

Regional USGS offices may maintain records concerning the
construction and logs of wells in the area. Records will be
available at the state, county, or local level if wells must be
registered by law. Sometimes health or environmental agencies
handling the collection of water quality data may also record these
well logs. A department of public works or water surveyor may also
maintain records containing well construction and well log
information.

4. Specific Discharges to Ground Water

Information concerning industrial locations, discharge
characteristics, and discharge permits will be available at the agency
level where regulation and enforcement are carried out. Regulatory
and health agencies may also maintain data regarding the locations of
sewage treatment plants, landfills, superfund sites, recharge basins,
and gasoline tanks. Planning agencies may have information concerning

possible   pollutant   sources,   usually   in   the   form   of   maps   or   summary
tables.   A   state   department   of   commerce   may   have   information   on
location   and   industrial   classification.   Data   relating   to   oil   spills
may   be   available   with   a   state   department   of   transportation.   State,
county   or   local   agencies   may   also   maintain   records   documenting   major
oil   or   chemical   spills.

5.   Land Use and Natural Resources Inventories

     Data   on   the   characteristics   of   the   surface   (other   than   specific
discharges)   must   be   included   if   investigations   are   to   be   made   of   the
relationships   between   activities   on   the   surface   and   ground   water
quality.   These   data   will   be   found   at   all   levels   of   jurisdiction   as
part   of   natural   resources   inventories.   Nationally,   the   USGS   maintains
two   extensive   data   bases   designed   for   a   comprehensive   inventory   of   the
nation's   natural   resources: LANDSAT   and   GIRAS   (Geographic   Information
Retrieval   and   Analysis   System,   of   Mitchell,   et   al.,   1977)   data   files.
GIRAS,   the   earlier   system,   contains   machine   readable   information   on
land   use   and   land   cover   (Anderson,   1976),   political   units,   hydrologic
units,   census   county   subdivision   of   Standard   Metropolitan   Statistical
Area   (SMSA)   tract,   and   federal/state   land   ownership.   Information   is
available   from   the   USGS   National   Mapping   Division   on   computer   tape   in
digital   format   at   scales   of   1:250,000   and   1:100,000.

     Natural   resources   and   land   use   activities   have   recently   been
inventoried   by   the   LANDSAT   program,   with   the   same   classification
system   as   used   by   GIRAS.   The   data   are   managed   exclusively   by   computer
and   are   immediately   available   for   the   years   since   1973   in   sections
that   view   the   same   area   on   the   surface   about   40   times   a   year.

     At   the   state   level   natural   resource   inventories   may   also   exist   in
machine   readable   form   at   departments   of   environmental   protection   and
transportation.   At   local   and   regional   levels   most   of   this   type   of
information   is   in   paper   files   and   may   not   be   comprehensive   (e.g.
budgetary   constraints   may   allow   only   partial   inventories   of   land
use).   Land   use   maps   are   usually   required   by   law,   and   aerial
photographs   are   always   available.   Since   most   of   these   data   will   be   in
paper   files   and   documents,   considerable   effort   is   required   to   digitize
and   transfer   the   information   into   a   computer.

6.   Zoning

     Information   concerning   zoning   and   zoning   changes   over   time   will
be   available   at   the   local   or   county   level   or   with   regional   planning
agencies.Other   information   can   be   found   in   local   planning   documents
(as   required   by   law),and   include   zoning,   tax,   and   real   estate   maps.
These   data   are   available   almost   exclusively   as   paper   documents.   In
general,   such   data   are   not   as   applicable   to   the   study   of   impacts   on
groundwater   as   the   natural   resources   inventories,   because   zoning   data
do   not   give   as   detailed   a   record   of   actual   site   conditions.   For
example,   a   large   area   of   land,   although   zoned   institutional,   may
actually   serve   many   land   uses   that   can   affect   ground   water,   such   as
residential,   open   space,   agricultural,   etc.

7. Population

Demographic data are available from the Bureau of the Census, which also maintains a file and software package called GBF/DIME (Geographically Based File/Dual Independent Map Encoding). GBF/DIME can "match" addresses with census blocks and tracts for areas of the country for which maps have been digitized; a "match" can sometimes be made between addresses and latitude/longitude locators. However, success in using GBF/DIME depends to a large extent on whether the local agencies have committed their resources to applying the software to the local geographic area.

8. Soils and Topography

Soil maps are available from the Soil Conservation Service. A computerized file describing the soil classifications used is also available. Topographic maps are available from the USGS.

Design of Data Base Structures and Schemas

A schema that is defined and loaded with data in the DBMS environment is based on a logical structure which reflects the relationships between variables. This structure may be strongly influenced by the structure of the data sets or data bases received from originating agencies. The structure will also depend on whether the MIS consists of one data base containing all of the required variables or on several related data bases. A well designed structure and schema will improve the efficiency of data retrieval and analysis. As described earlier, a DBMS will ordinarily allow the data base to be structured as a hierarchy, which is useful for many applications although other structures may also be used. The applicability of a hierarchical structure to ground water data is easily envisioned (see Appendix I). A particular well that has associated with it information concerning locations, construction, and hydrogeology may "own" records related in a hierarchical manner that contain sample dates and associated chemical data (Figure 2).In the same way, an industry identified by a number, name, location, and classification could be associated with a number of records containing permit, discharge, and compliance information. A hierarchical schema in the System 2000 DBMS can be found in Kaplan et. al. (1985). A data base that is to hold data concerning both industry and wells must relate them by some common descriptor. Locational information (such as a grid/coordinate system) could "own" records containing both industry and wells in that location. The industry and wells can, in turn, "own" records containing permit and water quality data (Figure 3).

Probably the most versatile structure for a ground water data base is relational. This data structure is organized into tables, and information in separate tables can be related by some common identifier. An example of a relational data base design is found in Appendix II.

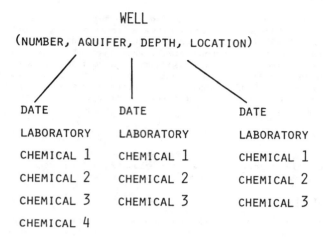

Figure 2.   Structure of a "Well Based" Hierarchical Data Base

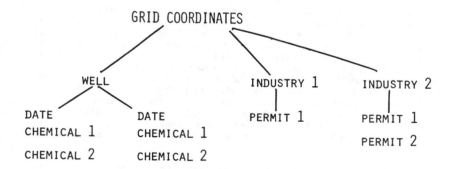

Figure 3.   Structure of a "Grid Based" Hierarchical Data Base

Once a logical structure for the data base has been determined, the schema must be built and defined in the language of the DBMS. The definition of the data base is made with knowledge and consideration of the form and formats of the data sets to be loaded into the data base. Room must be left for large numbers, and care must be taken to define character and integer variables properly. If the DBMS allows the schema to include line and column specifications in the schema definition, this information will be required prior to the definition of the data base. Data sets can usually be reformatted to match a predesigned and defined data base. The design and definition of a data base, should be dictated by the final use and application of the data base, the DBMS capabilities and language, and the structure and format of the available data sets.

Geographic Information Systems (GIS)

Concepts developed for a management information system can be extended by the perspective of the spatial relationships required in ground water analysis. By focusing on other types of data inferences regarding water quality can be extended to planning and management of surface activities that may contribute to aquifer contamination. The resulting system is a GIS that organizes data according to geographic location and facilitates spatial analysis.

Definition of Typical Spatial Locators/Types of Data

In the development of a GIS from diverse data sources, the initial technical problem is that different data will be located in different coordinate systems. Fortunately, in almost all cases spatial coordinate systems can be translated into other systems with existing computer software. Once acquired, this software may become part of the GIS. This section describes the types of coordinate systems likely to be encountered and the sources of transformation software.

1. Latitude/Longitude

The most common spatial locator used in data related to ground water is latitude/longitude. This system is most easily applied in the field by surveyors and, typically, well locations, discharge permits, and specific point sources of contamination will be located by latitude/longitude. All data associated with wells (such as construction logs and water quality parameters) will therefore also have a geographic locator. In general, any monitoring or regulatory compliance data collected for inventory or overview at discrete points as opposed to areal extent are located in this system.

2. Universal Transmercator Projection

Another common coordinate system is based on the universal transmercator projection (UTM). UTM projections slice the earth's surface into 60 north/south strips, 6 degrees wide, between 84° north latitude and 80° south latitude. A rectilinear grid system was developed for use with this projection so that any point can be

easily located. Any point on the map can be exactly located by
measuring (or estimating) its metric distance north and east of the
nearest reference lines and adding this amount to the reference
coordinates. This leads to the most important attribute of the UTM
projection grid system: by applying the Pythagorean relation to the
metric coordinates of two points, the distance in meters between them
can be calculated. Nonpoint source data are typically associated with
this system. Examples such as land use, zoning, topography, soil
types, and population densities can usually be found locally and
regionally in maps and associated tables. Municipal and county
governments, as well as regional offices of federal and state
agencies, are the best sources for these data. Data on specific
activities like agriculture or tourism can be found with such
agencies as the local agricultural extension service (soils
information) and the chamber of commerce (population patterns).
Unfortunately, management of this information is difficult since in
most cases it is not stored in computer memory but rather in paper
files.

3. Other Coordinate Systems

    Other projections and coordinate systems may be used to locate
spatial data in addition to geographic, universal transmercator, and
state coordinate systems. In any particular situation the agency or
group responsible for the data will likely be able to supply methods
to transform to any other coordinate system. The most important
example of the use of other coordinate systems is with the USGS's
program of LANDSAT data. The system used to assign coordinates to
units of information (pixels) depends on the section of the globe that
a particular LANDSAT image represents.

4. Address (GBF/DIME FILE)

    A large amount of detailed spatial information is located by
address. Any locally collected data; such as private well locations,
small discharges, recharge basin location, accidental impacts, minor
landfills, gasoline tanks, etc., will generally be located in this
manner. Unfortunately, any system of addresses is locally specific
and not easily translatable into the geographic or other coordinate
system. Although the U.S. Bureau of the Census maintains GBF/DIME
FILE software to translate addresses to specific census blocks and
nodes for any area of the country, association of the nodes with a
spatial coordinate system requires considerable manual digitization
and, therefore, varies with the particular area under study.

Choice of Spatial Locator

    A common locator must be specified before aggregation of diverse
data for comparative studies is begun. This will be the coordinate
system to receive translated data from all other systems, and the
choice should be based on several considerations:
    o scale of application and size of study area
    o locators and formats of available data
    o type of application (eg. modeling, statistics, mapping)
    o output and presentation

Although each situation will be unique, the following general recommendations are associated with the choice of a specific spatial locator.

1. Scale of Application

The larger the region of study, the more likely it is that the common locator of input data will be geographic (latitude/longitude), because the errors resulting from extrapolation in planar projections increase as the size of the region increases. Rectangular systems like UTM are adequate if the region spans no more than 6° of latitude or longitude, and software may exist to increase the useful span by several degrees if this is necessary in the particular region of study. (For example, New York State extends its state coordinate system beyond the normal span of UTM zone 18 in order to include Long Island).

2. Locators and Formats of Available Data

If much of the data is coded in one type of coordinate system it is probably worthwhile to make that the common locator; but other considerations such as format specifications of the DBMS may dominate if transformation requires a restructuring of a particular data base. This type of tradeoff centers on how much effort has gone into formatting some data sets and how much their reformatting would add to the total effort.

3. Type of Application

The type of analytical application of the GIS will help determine the choice of a common spatial locator because of requirements of different analytic techniques. Methods of spatial analysis such as interpolation by triangulation or by kriging, and numerical modeling may require data input in a specific form. In addition, mathematical manipulation involving certain coordinates on a small scale (i.e., within one degree of latitude square) may present precision problems on digital computers since many digits to the right of a decimal point must be maintained. When small distances are calculated in geographical coordinates, for example, problems may develop as the numbers are truncated or rounded during computation.

4. Output and Presentation

More programming is required for curved output projections than for flat projections of straight lines. The convenience of using flat projections should be balanced against the need for sophisticated presentation graphics depending on the purpose and resources of the user. Use of geographic coordinates requires some compensation for the earth's curvature whereas a grid system like the one associated with the UTM projection has minimal distortion in a plane.

Assembly of Diverse Spatial Data

The GIS goal of aggregation for comparison and analysis of diverse spatial data involves a consideration of the resolution level at the same time the common locator is determined. Resolution level means the areal extent of the smallest unit of spatial information in the GIS. This choice relates directly to the purposes of the user, especially the scale of the user's application, but can also be fixed by the data alone if the units of information in a particular data set are large.

Although coordinates of point information such as wells and point discharges can be represented exactly in computer memory, the coordinates of areal information such as land use, nonpoint pollution sources, soils groupings, and geologic units cannot be specified exactly. A scale of resolution must balance the increased information possible with smaller units against the ease of data manipulation and storage of larger units. In addition, the user's purpose may limit the amount of information necessary to perform analyses, and this will help to determine the resolution of the GIS. For example, resolution of a few feet is unnecessary when ground water travels several yards per month and the user wishes to focus on the seasonal effects of rainfall.

A major consideration in the design and use of the GIS is the shape of the smallest unit of data, i.e., whether the information will be in a polygonal or a grid format. The choice between systems is based on the detail of information possible with the polygon format and the ease of data management possible with the grid format.

Polygonal formats store the exact outline of a feature on the land surface by maintaining a digital record in terms of the coordinates of arcs and nodes of the boundary of the polygon. In addition to this information and the descriptive code of the feature being outlined, polygon format must store an indication of points inside and outside the polygon. The major advantage of this format is that a detailed record of transitions between areal features is stored in computer memory. A major disadvantage is that the amount of computer memory dedicated to storage can be large. In addition, errors in data manipulation such as overlay or simulation modeling increase when different polygon formats are compared. This is particularly true for overlay of different polygon systems since new polygons are created where boundaries do not coincide (Clarke, 1983; Franklin, 1983). Another disadvantage of polygon format is the time and effort necessary to transform paper files to computer polygon format, since this process entails manual interpretation of maps and manual input into the computer. However, certain applications may require the detail that can be achieved with polygons.

Grid format divides the land surface into a grid system of arbitrary size, depending on the purpose of the user. Computer memory stores the coordinates of each grid (usually the centerpoint) and the code of the feature the grid contains within its boundary. Overlay of information from different years is easily accomplished in a structure of this type, as years change simply through a change in feature code. Therefore, the grid system is better suited to a study of temporal changes than of spatial changes. Another advantage is that grid format facilitates transfer of paper file data to computer memory. Since information is obtained at regular intervals, mechanical techniques can replace manual digitization. A major disadvantage is that boundaries between features become obscured to the level of resolution of the grids. However, grids can be made sufficiently fine for most user needs. Another disadvantage of grid format is that by obscuring the actual boundaries on the land surface there will often be more than one feature within a particular grid. The problem is then proper assignment of a feature code to that grid. One solution used by the (USGS) is to assign the code of the feature at the center point to the entire grid, but for features of very small areal extent this may mask too much data to be acceptable. This is particularly true in ground water studies where small nonpoint sources can have a major effect on the ground water quality. Hence this biased technique may lead to an unusable data set. The problem is mitigated somewhat by designing a grid system with finer resolution and by assigning a code in a way that maintains information in addition to that at the centerpoint.

Summary and Conclusions

Decisions affecting ground water are increasingly being made based upon diverse data which exist at many different levels of government and in the private domain. The body of such data is growing quickly. Computer based systems are required to assemble and manipulate this information. Software are presently available in the public domain which address some of the information system and calculational needs of those involved with ground water data. An integrated package is needed to allow a user to easily assemble required information, and then to perform whatever statistical, graphical, and numerical modeling calculations are necessary. More private consulting firms than ever before are collecting data in digital form, and are using microcomputers to analyze this information. It would make the life of future geohydrologists considerably simpler if standards were arrived at for the storage and electronic transfer of ground water information. Based on the growing number of software packages available to the practicing geohydrologist, it would appear that consensus will soon be reached as to the data base management systems of preference, as well as those numerical techniques most favored and useful. The ability to perform more complex analyses on better quality data will soon allow for better engineering approaches to the monitoring and protection of our ground water resources.

APPENDIX I
DIVERSE GROUND WATER RELATED DATA

SITE/FACILITY DESCRIPTORS

1. Site descriptors

location (e.g. address; latitude/longitude)
wastes found on site
wastes injected
site owner
legal description
agency responsible for oversight/cleanup
number of wells on site
number of nearby offsite wells (and locations)
sources of onsite contamination (e.g. lagoons,
    storage tanks, etc)

2. RCRA Manifest Information

WELL DESCRIPTORS

1. Well location

latitude/longitude
FIPS county code
UTM coordinates
town, range, section, quarter
state grid system
site specific numbering system

2. Average depth to ground water

3. Aquifer code(s)

4. Water Quantities

pump rates
aquifer yield(s)
pumping schedule

5. Availability of geophysical log

6. Availability of well drillers log

7. Well Characteristics

pipe top elevation
date of construction
name of driller
treatments

8. Well type

     irrigation
     drainage
     industrial supply
     domestic supply
     municipal supply
     recharge
     monitoring
     other

9.   Well purpose (i.e. RCRA, Superfund, drinking water,
     injection)

10. Construction methods

     air rotary
     bored
     augered
     cable tool
     hydrologic rotary
     jetted
     air percussion
     reversed piston
     turbine
     other
     unknown

11. Casing materials (may need associated depths)

     PVC
     teflon
     ABS
     brick
     concrete
     copper
     steel (e.g. stainless)
     rock or stone
     other

12. Screen characteristics (may need associated depths)
     number of screens
     depth to screen(s)
     materials
             ABS
             brass
             galvanized iron
             wrought iron
             black iron
             TBC
             stainless steel
             teflon
             tile
             other

```
size of screen(s)
   width(s)
   slot size(s)
```

13. Well Status

```
        Abandoned
        Flowing
        Non-flowing
        Plugged
           depth of plug(s)
           type of plug(s) (packing)
```

GEOLOGIC/HYDROLOGIC DESCRIPTORS

1. Geologic/hydrologic

```
    aquifer depth(s)
    aquifer type(s)
       (include aquicludes/aquitards)
       description
       confined/unconfined
       permeability
       transmissivity
       porosity
       hydraulic conductivity
       depth to water table (may need date/time)
       subsurface stratigraphy
       lithology
```

2. Topography

```
        location of discharge and recharge areas
        surface water flow pattern(s)
```

3. Soils

```
        horizon
        depth
        group/type
           associated attributes
              permeability
              transmissivity
              porosity
              other
           other descriptive information
```

WATER LEVEL DATA

1. Date of measurement

2. Well identifier

3. Depth to water

    technique used (tape,acoustic,laser)
    height above mean sea level
    depth referenced to some other datum
    error associated with measurement

WATER QUALITY/SAMPLE DESCRIPTORS

1. Sample identifiers
    date of sample
    depth of sample
    name of collecting agency
    name of analyzing agency
    name of laboratory
    submitting agency code

2. Sampling Protocol

    number well volumes removed prior to sampling
    air lift pump
    submersible pump
    bailer

3. Sample Type

    grab
    split sample
    24, 12, 8 or 6 hour composite
    duplicate sample
    treatment

4. Analytic Method

    USGS Standards
    EPA Standards
    local agency standards
    instrument calibration
    detection limit (zero not sufficient)
    standard method code
    QA/QC CODE

5. Water Quality

    standard chemical parameter codes for constituents
    measurement unit
    value of measurement

SPATIAL DATA

1. Location of other regulated facilities (e.g., RCRA,
    SF, small quantity generators, ground water
    discharge permit holders)

2. Location of other wells

    public supply wells
    USGS monitoring wells
    local agency monitoring wells
    private wells

3. Population

4. Land use

    location ground water dischargers
    recharge basins
    area industry by SIC code
    agricultural areas
    pesticide usage
    ground cover
    highway network
    oil gas pipeline network

APPENDIX II
RELATIONAL DATA BASE

An environmental monitoring data base has been implemented at BNL on IBM microcomputers. The data base stores data associated with ground water and surface water samples. Data is entered to facilitate the production of tables for yearly environmental monitoring reports. The long range plan is to purchase additional microcomputers to allow data entry and retrieval at several work stations. The relational data base management system (DBMS) used is RBASE (Microrim Inc.).

This data base stores data associated with ground water and surface water samples. Samples are "logged in" and entered in the LOGIN table before chemical and radiological data is entered. Samples are identified as grab, proportional or composite sample by the TYPE column. Water level data and chemical data are stored in separate tables. Sample information can be linked with analytical data in any table because the columns LOCATION and LABID are held in common between tables.

There are several tables which store information which may be used in reports or files for further analyses. DWELLS is a table which stores descriptive information for stations such as coordinates, area, and well depth. Other tables in the data base include WLEVELS which stores water level data collected for wells and WCHEM which stores chemical data for all samples. The schemas for LOGIN, WLEVELS, WCHEM and DWELLS are presented below.

TABLE: LOGIN

Column definitions

# Name	Type	Length
1 LOCATION	TEXT	10 characters
2 LABID	TEXT	15 characters
3 TYPE	TEXT	4 characters
4 DATE	DATE	1 value(s)
5 PH	REAL	1 value(s)
6 COND	REAL	1 value(s)
7 TEMP	REAL	1 value(s)
8 DO	REAL	1 value(s)
9 LAB	TEXT	10 characters
10 METHOD	TEXT	5 characters
11 COMMENT	TEXT	40 characters
12 QA	TEXT	2 characters
16 FREQ	TEXT	2 characters
21 FLOW	REAL	1 value(s)

Table: WLEVELS

Column definitions

#	Name	Type	Length
1	LOCATION	TEXT	10 characters
2	LABID	TEXT	15 characters
3	CDATE	DATE	1 value(s)
4	WLEVEL	REAL	1 value(s)
5	ERROR	REAL	1 value(s)
6	HAMSL	REAL	1 value(s)
7	LAB	TEXT	10 characters
8	QA	TEXT	2 characters

Table: WCHEM

Column definitions

#	Name	Type	Length	Key
1	LOCATION	TEXT	10 characters	
2	LABID	TEXT	15 characters	
3	ANDATE	DATE	1 value(s)	
4	TYPE	TEXT	4 characters	
5	CHEMICAL	TEXT	12 characters	
6	VALUE	REAL	1 value(s)	
7	CERROR	REAL	1 value(s)	
8	LOADING	REAL	1 value(s)	
9	QA	TEXT	2 characters	

Table: DWELLS

Column definitions

#	Name	Type	Length
1	LOCATION	TEXT	10 characters
2	AREA	TEXT	20 characters
5	DESCR	TEXT	40 characters
6	PIPETOP	REAL	1 value(s)
7	DEPTH	REAL	1 value(s)
8	DIAMETER	REAL	1 value(s)
9	LAT	REAL	1 value(s)
10	LONG	REAL	1 value(s)
11	COORDX	REAL	1 value(s)
12	COORDY	REAL	1 value(s)
13	QA1	TEXT	12 characters
14	QA2	TEXT	12 characters
15	COMMENT	TEXT	20 characters

APPENDIX III
References

Anderson, J.R., Hardy, E.E., Roach, J.T., Witmer, R.E., 1976. A Land Use and Land Cover Classification System for use with Remote Sensor Data, U.S. Geological Survey, Professional Paper 964.

Clarke, K.C., 1983.   Polygon to Raster Conversion Error: A Comparative Analysis of Techniques, Seventeenth International Symposium on Remote Sensing of Environment, Ann Arbor, Michigan, May 9-13, 1983.

Franklin, W.R., 1983. "A Simplified Map Overlay Algorithm," Harvard Computer Graphics Conference.

Kaplan, E., J. Naidu, M. Hauptmann and A. Meinhold, "Guidebook for the Assembly and Use of Diverse Ground Water Data," prepared for the U.S.E.P.A., BNL Report 37356, April 1985.

Marble, D.F., Calkins, H.W., Penquet, D.J., 1984. Basic reading in Geographic Information Systems, SPAD Systems, Ltd., Williamsville, NY.

Martin, J., 1977.   Computer Data Base Organization, Prentice Hall, Englewood Cliffs, NJ.

Mitchell, W.B., Guptill, S.C., Anderson, K.E., Fegeas, R.G., Hallam, C.A., 1977. GIRAS: A Geographic Information Retrieval and Analysis System for Handling Land Use and Land Cover Data, U.S. Geological Survey, Professional Paper 1059.

Techniques for Delineating Subsurface Organic Contamination: A Case Study

Ann M. Pitchford*, Aldo T. Mazzella*, and Edward Heyse**

Introduction

Selection of techniques to detect and map subsurface organic contamination depends on the type of source, contaminantion, and hydrogeologic setting. Ideally, an investigator, manager, or enforcement official could choose mapping techniques such as soil gas and geophysical measurements, coring, or installing wells to delineate the contamination in a sequence of progressively greater precision and expense. This paper presents the results of applying several of the above techniques to a service station with a leak from a gasoline storage tank. The merits of and information gained from geophysical and soil gas measurements are compared to the results from analysis of water samples at existing wells.

Background Geology and Hydrology

The study site is located in the West, in an arid structural basin bounded by fault block mountains. Unconsolidated alluvium of Tertiary and Quaternary age fills the basin to depths of several thousand meters. Near the surface, fine-grained, gypsiferous sediments with low to moderate permeability dominate. Average precipitation is approximately 25 centimeters (cm) per year, and evapotranspiration is high with a small net inflow of water to the ground-water system. Most of the ground water originates in the surrounding mountains and moves downgradient into the basin along alluvial channels. At the study site, the ground water in the uppermost aquifer is at a depth of 2 to 7 meters (m), is highly mineralized, and is not used for drinking or irrigation. Recharge from watering of lawns may be significant in describing localized features of the water table during some seasons of the year.

Site Description

In 1981, discrepancies were noted in fuel inventories at the service station; 380,000 to 570,000 liters (100,000 to 150,000 gallons) of gasoline are believed to have leaked into the ground water. Test borings showed a layer of free gasoline over a meter thick in places on top of the water table. Some of the gasoline was removed by a recovery well before this study, but thicknesses of 0.7m remain. Strong gasoline odors have been noticed in the service station area and from a sewage pumping station across the street.

*Environmental Monitoring Systems Laboratory, U. S. EPA, P. O. Box 15027, Las Vegas, NV 89114
**Air Force Engineering Services Center, Tyndall Air Force Base, FL 32403

Objective and Approach

The objective of this paper is to qualitatively compare the results
from ground-water sampling with the results from several other techniques, to see what information each provides on the lateral extent of
the contamination.

The study site is shown in Figure 1. In a previous investigation,
thirteen wells had been installed. Measurement techniques initially
considered for use in this study included d. c. resistivity, soil gas,
soil cores, and well installation. Soil cores and installation of
additional wells were not considered further due to cost and because of
the many wells already present. Ground-water and floating product
samples were collected from the existing wells and analyzed and resistivity and soil gas measurements were made during the summer of 1985.

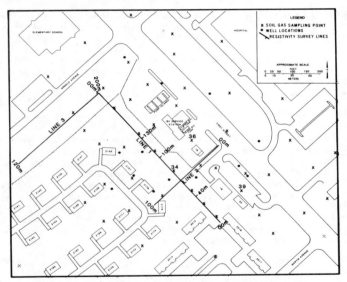

Figure 1. Study Site and Measurement Locations

The measurements performed at the gasoline station are described in
Table 1. Lockheed Engineering and Management Services Co., Inc. had the
lead responsibility for the field measurements, and performed the geophysical measurements and water sampling. Under subcontract to Lockheed, Tracer Research Corporation performed the soil gas measurements
(Lappala and Thompson, 1984; Marrin, 1985), and Western Technologies
Laboratory analyzed the water samples using standard EPA methods.
Measurement locations shown in Figure 1 were based on the area of contamination indicated by the previous investigation and did not have the
benefit of the information from the soil gas measurements or more recent
ground-water samples. These measurement locations are a compromise
among accessibility, proximity to the contamination, and presence of

TABLE 1.   MEASUREMENTS PERFORMED
===

Measurement; Instrument	Purpose
D. C. Resistivity, pole-dipole configuration; Bison Offset Sounding System	determine lateral changes in electrical resistivity as a function of depth
Soil gas; Varian 330 Gas Chromatograph and Spectra Physics 4270 Computing Integrator	determine concentrations of organic compounds in soil gas
Water Samples; pump, tubing	determine concentrations of organic and inorganic compounds in ground water and floating product
Water Level, Depth of Floating Product; measuring tape, indicator paste	determine water level and depth and thickness of floating layer of gasoline in wells

===

underground utilities. Pole-dipole resistivity measurements were made
with 1.7-meter and 5-meter dipoles along three transects, at locations
chosen to avoid fences, the asphalt apron for the gas station, and
streets and parking areas. The transects, labelled 1, 2, and 3, are
200, 100, and 120 meters long respectively. Soil gas was collected
from a total of 47 sampling points near the gas station, nine of which
were below paved surfaces and required the drilling of holes through
concrete and asphalt. Measurements of water level and thickness of
product were also made when the water and floating product samples were
collected at each of the thirteen wells.

Ground-Water Sampling Results

     Contour maps of concentrations of xylene and toluene dissolved in
ground water, and of concentrations of benzene, xylene, and toluene in
floating product all show the same general pattern; high concentrations
in a roughly triangular pattern surrounding the gas station. Figure 2,
showing concentrations of toluene in ground water, is typical. The
areas of high concentration defined by the ground-water sample analysis
are defined as contaminated for purposes of comparison with the other
techniques.

Geophysical Results

     Preliminary analysis of the geophysical data did not clearly reveal
the location of the hydrocarbon contamination. Figures 3 and 4 compare
the apparent resistivity profiles for typical segments along the tran-

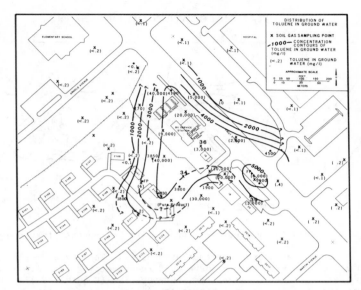

Figure 2. Concentrations of toluene in ground water.

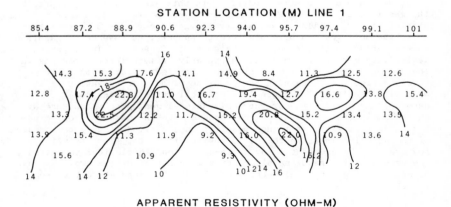

Figure 3. Resistivity pseudosection results for contaminated area.

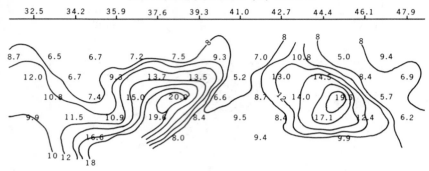

STATION LOCATION (M) LINE 3

APPARENT RESISTIVITY (OHM-M)

Figure 4.   Resistivity pseudosection results for uncontaminated area.

sects in contaminated and uncontaminated areas. This type display is
termed a pseudosection and was developed by converting pole-dipole data
to dipole-dipole data to simplify interpretation (Coggin, 1971). Both
segments show variable resistivities laterally and as a function of
depth.   The location of the gasoline plume as determined from Figure 2
extends on Line 1 in the area of station locations 40 to 130. At a
location about 5m from Line 1, station 90, the depth to the water table
was 2m and a gasoline layer of 0.4m thick was identified from a boring
made when the wells were installed in 1984. The apparent resistivity
contours in Figure 3 in the area of station locations 87 to 95 suggest
a two-dimensional resistive body lies in this area. This may represent
the gasoline layer. The data shown in Figure 4 are in an uncontaminated
area of Line 3. While no wells were located along Line 3 to confirm
this, the soil gas measurements seem to support the absence of gasoline
contamination in this area. (See Figures 5, 6, and 7.) The apparent
resistivity variations in Figure 4 probably reflect lateral changes in
the geology. The resistivity variations in Figure 4 are similar to
those in Figure 3 containing the gasoline plume. It appears that at
this site the lateral changes in electrical resistivities due to
naturally occurring two- or three-dimensional geological structure are
on the same order of magnitude as those due to the gasoline contamination. At this site, it is not possible to clearly delineate the presence
of the gasoline layer in the d. c. resistivity data. The resistive
features may be due to two- or three-dimensional geological variations,
the presence of a gasoline plume, or a combination of the two. The use
of additional geophysical methods may be able to separate out these
effects.

Soil Gas Results

Soil gas samples were analyzed for the presence of methane, benzene, toluene, o-xylene, 1,1,2- trichloro trifluoro-ethane (F113),
1,1,1- trichloroethane (TCA), trichloro ethylene (TCE), perchloroethylene (PCE), ethylbenzene/ xylenes, and total petroleum hydrocarbons.
The ethylbenzenes/ xylenes category included three aromatic hydrocarbons
which are not separated by the chromatographic column. F-113, TCA, TCE,
and PCE were not found in significant quantities. For comparison with
ground-water data, Figures 5, 6, and 7 show concentrations of the major
components of gasoline; toluene, benzene, and xylene, respectively.

Discussion

The soil gas data (Figures 5, 6, and 7) appear to delineate the
contaminated zones indicated by the water samples rather well, even
though the ground water and soil gas measurements were not colocated.
The patterns developed from contouring benzene, xylene, and toluene soil
gas data are very similar. The boundary of the contamination appears to
be fairly sharp, with steep gradients in the contamination over short
distances. Additionally, it is interesting to note that in each case
for ground water and soil gas, the highest concentrations occur at the
points of the triangle, not near the center where the leak occurred.
These locations are 1) near the pump islands; 2) along the street,
southeast of the station, and also across the street; and 3) near the
residences, southwest of the station. Very low values occur at locations where high values would be expected based on the local hydraulic
gradient; i.e., points 34, 36, and 39. Possible reasons for these
characteristics include one or more of the following:

  o movement of fuel in naturally-occurring zones of high permea-
    bility,
  o effects of local ground-water recharge,
  o presence of additional leaks,
  o effects of pumping during the recovery program, or
  o effects of utility corridors.

Figure 8 shows the underground utilities in the vicinity of the gas
station, along with the areas of highest concentration. Note the
presence of water lines with the potential for connecting the leak to
the areas of high concentrations both northwest and southeast of the
station along the street, and across the street. The presence of water
lines also offers a possible explanation for the high values to the
southwest in combination with the low values measured at point 34.

Movement of organic contamination along utility corridors has been
observed before (Nadeau, 1985). If this is a likely possibility, then
the methods of choosing sampling locations should be re-evaluated. In
this case, sampling perpendicular to known underground utilities should
be considered to delineate the contaminated zones.

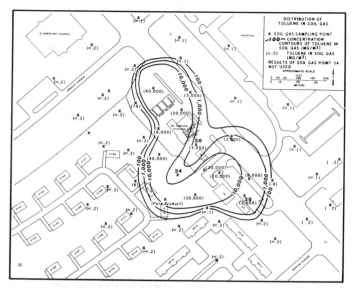

Figure 5. Concentrations of toluene in soil gas.

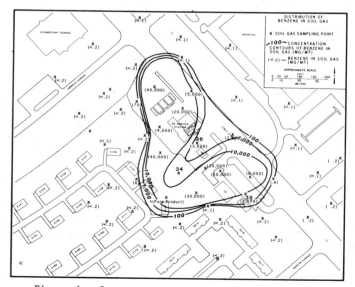

Figure 6. Concentrations of benzene in soil gas.

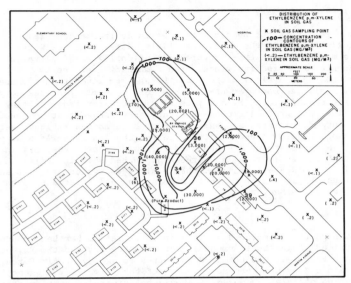

Figure 7. Concentrations of xylene in soil gas.

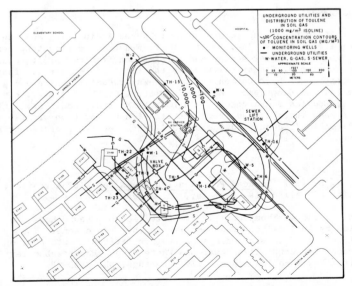

Figure 8. Underground utilities near service station
and areas of high soil gas and ground water contamination.

108   CONTAMINATED GROUND WATER

Conclusions and Summary

At this site, it is not possible to clearly delineate the presence of
the gasoline plume in the d. c. resistivity data. The resistive
features may be due to two- or three-dimensional geological variations,
the presence of a gasoline plume, or a combination of the two. The use
of additional geophysical methods may be able to separate out these
effects. Soil gas measurements delineated a contaminated area very
similar to that depicted by analysis of ground water from existing
wells. The soil gas and ground-water analyses showed that concentrations
in the contaminated area are variable with high concentrations near the
leading edge and low concentrations nearest the leak. While a number
of possible explanations are proposed for this, one of the more intrigu-
ing is that the fuel travelled through disturbed soil in utility corridors. If this is a possibility, then a high density coverage of samples
along lines perpendicular to known underground utilities should be used
to map the contamination.

Notice

    Although the research described in this article has been conducted
by the United States Environmental Protection Agency and the United
States Air Force, it has not been subjected to Agency or Air Force review
and therefore does not necessarily reflect the views of the Agency or Air
Force and no official endorsement should be inferred. Mention of trade
names or commercial products does not constitute endorsement or recom-
mendation for use.

Acknowledgement

    The field study was conducted by Lockheed Engineering and Manage-
ment Services Company, Inc. for the U. S. EPA under contract number
68-03-3245. This paper summarizes information from two draft reports,
"Interpretation of Electrical Geophysical Data" by Douglas J. LaBrecque,
and "Detection of Hydrocarbon and Ground-Water Contamination using
Geochemical and Geophysical Methods" by Steve J. Nacht, B. Virginia
Nicholas, Douglas J. LaBrecque, and Donn L. Marrin. Additional information has been incorporated from "Installation Restoration Program Phase
II Confirmation/ Quantification Stage 1 Draft Report" prepared by Dames
and Moore.

References

Coggon, J. H. 1971. Induced Polarization Anomalies. Dissertation, Uni-
    versity of California, Berkeley, CA, Appendix IIA.

Lappala, E. and G. Thompson. 1984. Detection of Ground-Water Contami-
    nation by Shallow Soil Gas Sampling in the Vadose Zone: Theory and
    Applications. in Proceedings of the 5th National Conference on Man-
    agement of Uncontrolled Hazardous Waste Sites, Hazardous Materials
    Control Research Institute, Silver Springs, MD, p 20-28.

Marrin, Donn L. 1985. "Delineation of Gasoline Hydrocarbons in Ground Water by Soil Gas Analysis," in Proceedings of the Hazardous Materials Management Conference/West '85, Tower Conference Management Company: Wheaton, IL, p 112-119.

Nadeau, Royal. 1985. Personal communication. U. S. EPA Environmental Response Team, Edison, NJ.

Some Problems in the Engineering of Ground Water Cleanup

John M. Armstrong[*]

Abstract:

The problems facing engineers, often civil engineers, in addressing ground water contamination are both exotic and mundane, but all are critical to success. The paper provides a brief review of selected problems encountered that engineers must consider in order to conduct efficient project engineering.

Introduction:

The level of marketing, analysis, study, design, and litigation in ground water contamination has been growing exponentially over the last 7-8 years. Never have there been so many conferences, seminars, short courses, and workshops as are available today in the general area of ground water contamination and cleanup.

The engineer, facing the realities of finding a solution to ground water cleanup, has a body of literature to review that is growing enormously every month. This is particularly true with respect to the development of models for predicting ground water flow and contaminant transport. During the last year I estimate that there were at least 350 papers, articles, and reports available for the engineer to use in dealing with ground water problems. (Ground water models are wonderful, if we can have some reasonable assurances that they are at least fairly accurate and that they apply to the specific and real environment being dealt with at the time. These are two criteria that are rarely achieved.)

In ground water contamination problems the civil engineer finds a situation similar to most environmental cleanup problems, i.e., the extent of the problem must be assessed, the alternatives for cleanup (or management) of the contamination must be identified and evaluated, and the best system then designed and implemented.

In ground water contamination, as compared to surface water problems, the complexity of assessing the nature and extent of the problem is significantly greater. The added dimension of the soil or rock structure through which the ground water flows is variable laterally and vertically, and is almost always non-homogeneous. Models sometimes can give us a clue to where the contamination is going and how fast, but the difficulty is that often as not we don't know how much

[*]Senior Partner, The Traverse Group, Inc., 2480 Gale Road, Ann Arbor, MI 48105.

contamination was initially released, and often we don't know when, or exactly where. In addition, the exact nature and effects of biological and chemical phenomena active in the subsurface environment are often unknown.

This brings us to the main objective of this brief paper--to outline some of the realities that civil engineers will meet when they finally have to go out into the field (or forest) and tell their client where the ground water contaminant plume is and how intense it is. This, of course, must be done before any real progress can be made on determining which technical approach will work to clean and/or contain the contamination. In addressing ground water contamination the engineer will need to consider several steps that he or she must carry out on almost all engineering projects but specifically in dealing with ground water:

   o assessment of the site and extent of the contamination,
   o identification and analysis of alternatives for cleanup,
   o design of the selected cleanup,
   o construction of the cleanup system, and
   o monitoring the performance of the system. ·

In all five steps the civil engineer will find that ground water cleanup is one of the most interdisciplinary problems that can be undertaken. The engineer will need inputs from:

Chemists - to perform analysis of the soil/water samples
Biologists (sometimes) - to evaluate degradation potential of
     contaminants
Geologists - soil and geologic structure of the contaminated
     system
Technicians - to take the soil and water samples
Lawyers (sometimes) - to make the engineering more difficult
Mechanical and chemical engineers - to help design and build
     certain treatment equipment.

Thus in the opinion of myself and many engineers the ground water contamination problem is the most challenging, most interesting, and most difficult of all of the tasks facing the civil engineer.

To fulfill my charge from the ASCE in what I believe to be the most interesting way, I am going to devote the remainder of the paper to brief, separate discussions of some of the most challenging problems the ground water engineer will face as he or she arrives at the contamination site, eager and ready to help the world become a better place. Some of these problems are broad and sophisticated, while others appear mundane but are still critical to project success. This list is by no means complete. I will cover a number of issues here, from the constraints of ligitation and legal actions, to the practical problems of iron precipitate (familiar to many civil engineers, but now in a somewhat different context).

1. Ground Water Contamination "Plumes"

Finding out how far the released or spilled contaminants have
gone in the ground water is a major task in itself. Assuming the
spill has reached the ground water, there is usually a two-phase
character: (a) the presence and movement of the original spilled
material, and (b) the solution of some (or all) of the original spill
material into the water. The first thing a client is worried about is
whether or not he or she is the only responsible party or are there
others who may have contaminated the ground water prior to the
release. If so, the other parties may have to share the liability (or
"hold hands" with the client as lawyers describe it). This is particularly difficult if the client's location is in an industrial area
with many other potential release sources upgradient from the client's
property. The intermingling of a variety of plumes will probably make
the issue of distinct plume definition almost impossible to resolve
unless the potential contaminators are using or producing materials
that are significantly different in chemical character. The engineer
often is faced with the problem of being unable to state with
certainty how much of the contaminants in the aquifer is the client's
and which can be credited to other parties. Sophisticated chemical
spectral analysis can sometimes be used to discriminate constituents
from various sources. However, often as not "weathering" effects or
biochemical actions in the aquifer media may substantially alter the
contaminant's appearance and behavior so that differentiation becomes
impossible, particularly in formal litigation proceedings where
scientific "expertise" is used to split fine hairs on almost everything that is presented.

This brings us to another, but closely related, problem category
that often is the most controlling force in conducting ground water
engineering projects: these are legal actions and liability questions. The existence of litigation in ground water contamination
cases is becoming more and more prevalent and will continue to increase over time.

2. Litigation Constraints

The presence of litigation, of course, is not unique in the
engineering profession, but in ground water projects certain specific
constraints arise that affect the technical approach that might be
employed to cleanup or contain the plume of contaminantion carried by
the ground water through the aquifer.

For example, from a liability standpoint, or in regard to ongoing
legal proceedings against a client, it may be in the best legal
interest of the client to immediately block movement of the contaminant plume from going beyond the client's property line or if already
beyond, to prevent any further contamination from leaving the property. This can be done by designing and installing an interceptor
well system across the plume on the client's property to serve as a
"hydraulic wall". Such a system may or may not be effective in

removing large volumes of contaminants (or "purging" them) but may make perfect sense from a legal or liability standpoint.   In fact the design of the system from an engineering view may dictate a flow rate that will both capture the plume and also purge significant amounts of water.   However, an interceptor system designed just to block the plume from leaving may have far less flow requirements (and thus less costly), but still be legally effective.   Closure of the ground water potential lines by an interceptor pumping system may not mean closure of the litigation but often it can go a long ways towards helping litigation to be resolved or mitigated.

Often, motivation other than fact and engineering science lead to decisions regarding plans of action or policy.   Enforcement agencies often have policy guidelines on which they rely rigidly.   There is reluctance to allow exception in unique situations for fear of setting precedents.   This is the reason why so many sites under litigation are excavated and the soils landfilled, when the soils might have been easily biodegraded, or land farmed.   As much or more money may be spent collecting evidence to fight an unreasonable enforcement action as could be spent cleaning up an aquifer by natural or enhanced in-situ or land-farmed biodegradation.

Settlement agreements written by lawyers without input from engineers may lock the responsible parties into action plans that are not cost-effective or efficient.   An Air Force base signed an agreement with a regulatory agency stating that they would use activated carbon to clean up their ground water.   They are presently consuming $1,000 per day in activated carbon.   To use other treatment technologies which may be more efficient would involve reopening the litigation action in order to change the wording of the settlement agreement.

Another extremely difficult problem for the ground water engineer, related to legal liability problems, is the situation where a client wants to know if a site is contaminated before he buys or develops it. (If he buys without prior examination, he also buys the liability of the contamination if discovered later.)   In this case the engineer doesn't know where to look or what to look for.   A minute of thought will bring the engineer not previously faced with this problem to the next question which is how much of an examination of the site is enough to give the client a reasonable level of assurance? What is reasonable and how to attain it?

For even more practical examples of engineering realities consider the following, which are taken from actual projects and will be familiar to the experienced engineer.

3.   Taking Samples

One of the most important and also controversial issues in the ground water project engineer's bag of problems is the question of properly taking soil and water samples.   The size of the water sample, the method of obtaining the sample, how it is transported, and how it is analyzed in the laboratory are the subject of constant professional and legal discussion; the latter usually in court in litigation cases.

Unfortunately, there are no standard methods that yet reflect the consensus of practicing professionals and the reality of actual field work. The U.S. Environmental Protection Agency is trying to establish how samples should be taken and analyzed, but unfortunately they have not yet fully developed the scientific justification for all of their "guidelines". The engineer will have to spend a considerable amount of time determining what sampling approach is suitable and acceptable for his or her project objectives (technical and legal). The U.S. E.P.A. criteria and guidelines are generally being used, but they are by no means generally agreed upon by ground water professionals who have to work in the field.

4.   Precipitated Solids in Water Pumping

In treating water pumped from the contaminated aquifer (for intercept or purging) a frequent engineering problem is encountered with precipitated solids. This causes problems in the delivery and treatment systems, e.g., carbon adsorption. We will devote more discussion on this matter because it is pervasive in the field and it is an example of a conventional problem but yet plagues the speciality engineer. It is a problem that can spoil the best laid, technically efficient ground water cleanup plans.

For example, iron precipitate and other solids can build up in the activated carbon tanks at a rapid rate. The solids accumulation causes the inlet pressure of the carbon tanks to rise. The back pressure exerted on the pumps in the well fields steadily increases as the quantity of solids in the carbon tanks increases. The efficiency of pumping is steadily decreased with the increasing pressure. The well pumps may be incapable of meeting their design flow rates at a carbon inlet pressure of greater than say 45 psi. Also, the carbon tanks may be rated to withstand only 70 psi. The pipes and valves may break at pressures above 70 psi.

The precipitated solids can build up in the plumbing throughout the ground water pumping and treatment system. The solids reduce the inner diameters of the piping. This creates greater friction headloss and also reduces the efficiency of the well pumps. For example in a project at a U.S. Coast Guard Air Station in Michigan there is 60 feet of rubber hose, 2000 feet of buried 6-inch pipe, 400 feet of 4-inch pipe, and over 3000 feet of 2-inch pipe transporting water from the well fields to the activated carbon tanks and air stripper. Within the first 80 days of operation, a layer of orange solid precipitate approximately one-half to three quarters of an inch thick lined the inner circumference of these pipes and hoses. This system has been set up to run indefinitely, but the carrying capacity of these water lines may become negligible within a short time if nothing is done to impede the rate of solids accumulation.

On the microscale, iron often exists in the pristine aquifer in the forms of ferrous hydroxide, ferrous carbonate, and ferrous sulphide. All of these compounds exist in an equilibrium solution with free ferrous iron ions. The concentration of ions in solution is a function of the temperature, pH, and oxidation-reduction potential of the solution. At a neutral pH, in the presence of dissolved oxygen, most of the free iron ions in solution would be in the form of $Fe2+$. This concentration would be small relative to the concentration of insoluble ferrous compounds.

In the contaminated zone of the aquifer, iron bacteria of the species <u>Gallionella</u> utilize the ferrous iron as a energy source. The oxidation reaction of ferrous iron to ferric iron yields an electron which these chemoautrotrophic microorganisms can utilize in breaking the carbon bonds of carbon dioxide.

Ferrous iron is leached from the solid form into solution in order to replace the converted ferrous iron ions. The resultant concentraton of ferric ($Fe3+$) iron in solution is very high relative to that of pristine ground waters. Ferric iron is highly reactive due to its unstable valence state.

As the ground water enters the pumping wells, it is mixed with ground waters from other regions of the aquifer with higher concentrations of dissolved oxygen. The highly reactive ferric iron is precipitated out of solution as insoluble ferric oxyhydroxide.

The iron bacteria <u>Gallionella</u> are filamentous kidney-shaped microorganisms. The bacteria adhere to the inner walls of the piping system, valves, and fittings. They also accumulate at the top of the activated carbon. The tentacles of the bacteria intertwine forming a web and holding the bacteria together in a colony. The web of bacterial tentacles captures suspended solids in the solution such as the ferric oxyhydroxic precipitate. The characteristic orange color of the resultant sludge is due to the oxygenated iron.

Physically, there are several possible solutions to the solids accumulation problem. However, most require extensive capital investments and operational costs. At present, the solution operating under a program of periodic maintenance involving acid injection and backwashing of the carbon tanks.

5.   Time Required to Accomplish Cleanup

In determining the most cost-effective means of treatment and in predicting total project costs, it is important to know the duration of the project in order to amortize capital costs over the length of the project. Time to cleanup is, of course, important in litigation cases.

As yet, there is no reliable method of predicting the rate at which an aquifer will be reclaimed if contaminated water is pumped and treated at a given flow rate. The site specific characteristics of the soils, types of contaminants present, as well as the physiochemical conditions of the ground water, and the species of microorganisms present all interact to determine the rate at which an aquifer may be reclaimed.

Some of the important parameters are:

Soils
     Geology
     Hydraulic conductivity
     Adsorption capacity
     Ground water velocity

Contaminants
     Volatility
     Recalcitrance
     Reactivity
     Toxicity
     Mobility in soils

Physiochemical conditions of ground water
     pH
     Buffering capacity
     Temperature
     Dissolved oxygen concentration
     Nutrient concentrations
     Trace minerals concentrations

Microorganism
     Species
     Adaptability
     Resistance
     Metabolic rate
     Aerobic vs anaerobic

Besides the above technical factors that determine the time required for cleanup, the other most important factor is the popular question of "how clean is clean," or at what level of remaining contamination is the aquifer called clean?

This question may never be answered. One thing is certain, however, that in most ground water contamination cases the level of residual contamination will never be zero. In fact the term zero has no real meaning (e.g., what level is detectable, or non-detectable?).

6.   Lack of Data About the Man-Made Features of the Site

Many sites do not have accurate utilities diagrams describing what is buried below grade. In drilling to obtain soil and water samples or in trenching to lay pipelines this can be of serious concern. The main concern is in the danger posed to the equipment operators where there are unknown electrical and gas lines. Another concern is the cost of repairing damaged utility lines and the interruption to operations when power, light, phone, computer, or water lines are cut off.

For example, at a recent large ground water project, there were many unmarked underground electrical and communication cables. During the installation of the main interdiction pipeline, work was stopped for several hours when an unidentified electrical cable was encountered by the backhoe. Work was not resumed until it was assured that there was no current passing through the cable.

The project of winterizing an onsite purge field was held up for several hours while the owner searched for documents specifying the location of a high frequency radio antenna cables. Meanwhile, construction workers onsite sat idle.

During the installation of a water line to a ground water treatment facility, the backhoe operator hit a 1-inch temporary water line to an office trailer. As there was no utility diagram of the site, no one onsite knew where the valve to close the line was located. Flooding occurred before the valve was located. While repairing the water line, the backhoe hit the sewer line and broke the 8-inch clay main pipe leaving the site. As there was no utilities diagram, the owner assumed the responsibility for the repair bills which were $3500.

These kinds of incidents may seem mundane, but for project engineers they are critical.

7.   Public Perspective of Project

When sampling ground water for potentially hazardous chemicals, it is important for workers to wear protective gloves and respirators. In areas where the hazard level is undetermined, workers should wear a full protective suit complete with an air tank for breathing. When working in residential areas, this may cause the residents a great deal of alarm. Some project engineers have been hampered by reaction to onsite work.

In the case of the Love Canal, the most highly contaminated areas were fenced off for cleanup. Residents on just the other side of the fence were greatly disturbed to see workers in full protective clothing with enclosed air supplies working 50 feet away from their homes. Meanwhile, lawyers, politicians, and scientists were insisting to these residents that they were in no danger.

When lines of communication are not kept open between clients, contractors, government agencies, and the local community, misperceptions often occur. The resulting alarm can seriously hamper effective mitigation of the contamination.

8. Weather Problems in the Field

Ground water contamination does not always occur in mild cli-
mates. Some sites can be bitterly cold. Subzero temperatures are not
uncommon. At several sites where our firm has worked, one hundred and
fifty three inches of snow fell in each winter of 1985 and 1986.
Working outdoors with ground water is often difficult in such extreme
conditions.

All plumbing has to be buried below the frost line. It is not
feasible to run water through rubber hose above ground in the winter.
If flows do not cease the operation can be maintained, but inevitably
pumps will fail.

Taking water samples in the bitter cold is a difficult job.
Wells placed far from plowed roads must be hiked to on snow shoes.
The sample technician must carry well sampling equipment on a back-
pack. Gloves must be removed to place septa and to cap bottles,
increasing the danger of frostbite.

Pumping fuel from recovery wells through rubber hose is another
difficult task. At the start of each pumping excursion, the pump must
be thawed out with warm water carried from the nearest building. At
the U.S. Coast Guard project, the fuel pump was mounted on a sled to
be taken from well to well across the snow.

These are small but critical types of problems that can determine
project success or failure.

9. Site Realities as Constraints

Work at a site may be hindered by the presence of buildings,
parking lots, roadways, railroads, or other commercial developments.
It is not feasible to go into a shopping center and begin excavating
contaminated soil because a nearby gas station had a spill.

Foundations that are subject to settling are of prime concern
when purging from nearby wells. The reduction in the level of the
water table may cause a small amount of subsidence of the soil
surface. This subsidence may cause building foundations to settle
and walls to crack. Soils testing performed on the soils underlying
the U.S. Coast Guard's Hangar Administration Building showed that a
drawdown of three feet in the water table surface would lead to a
surface subsidence of one-fourth inch. Pumping flow rates from nearby
wells are carefully controlled to limit the drawdown.

It is difficult to accurately describe the subsurface geology of
a site. Educated hypotheses may be accepted and relied on based on
soil borings, radar, sonar, resistivity, and seismic surveys. However, all these data compiled do still not guarantee an accurate
analysis of what lies below the site. The data gathered by these
methods are subject to interpretation. Working where buildings cover

the original spill is a problem. Soil borings and geologic surveys, although sometimes possible, are difficult at best through building floors. The data may be unreliable as well, especially where underground utilities are nearby.

10. Political Effects

Politicians often want to get involved in environmental actions to improve their public image, even though their technical knowledge may be very limited. Political pressure applied to a decision may lead to hasty, and inefficient decisions. In a recent ground water case an enforcement agency reacted to a ground water case without a technical knowlege of the actual severity of the case. Political pressure lead to a lawsuit between the owner and the agency with very little chance for a peaceful negotiated settlement for several years.

The news media and concerned citizens groups seeking information to write about and concern themselves with will often make a larger issue out of a case than the case deserves. A lack of information does not stop these individuals from drawing their conclusions. This type of publicity creates undue concern on the part of the local homeowners in the vicinity of a site. In one case of ground water contamination, detrimental publicity lowered the value of the homes in the ajoining neighborhoods even though residents were provided permanent city water in place of individual wells. From a realtor's point of view, the homes are actually worth more now that they are hooked to city water. However, the sale prices are still devalued due to the publicity over the contamination in spite of the fact that there is no possible route of exposure.

While some of these problems are not directly involved with the engineering design work they often can indirectly affect the application of good engineering practice.

11. Improved Technologies

It is often the case that new, improved technologies become available to the engineer that allow more effective and less costly solutions to the problems of concern. This is especially true in relatively new fields, such as ground water contamination cleanup. Keeping pace with changing technology, although difficult to the busy engineer, is extremely important.

A good example of this in the field of ground water contamination is biodegradation. Traditionally, cleanup programs have followed fairly standardized pump-and-treat approaches. While often effective, such systems can be inappropriate when less costly and more efficient systems that exploit the natural biodegradation potential of the contaminants are available. This is sometimes as problem when hard-pressed regulatory agencies rely on what they believe are proven, conventional pump-and-treat approaches. Often these approaches will not accomplish anything but to spend a great deal of money.

One of the phenomena reducing the effectiveness of pump-and-treat systems is the tendency of some types of contaminants to remain trapped in the capillary zone immediately above the water table. These contaminants exist in the form of "microspheres" of product that are entrapped in interstitial pore spaces and/or adhered to soil particles. No amount of flushing or pumping can completely remove contaminants in this state, so that a continuing "source" remains below ground in the area of and downgradient of the spill.

Another effective (and often less expensive) way to remove this "source" is to allow the native bacteria, already present in the subsurface, to degrade the original product to harmless materials, such as carbon dioxide. This method is suitable with many hydrocarbon materials, such as fuels or some solvents. Aerobic (oxygen-using) bacteria are by far the fastest in consuming hydrocarbon contaminants. Aerobic degradation can be greatly enhanced by the additon of oxygen and (sometimes) nutrients and minerals. Oxygen addition in the form of hydrogen peroxide is one method used. However, such systems must be carefully designed to provide enough oxygen to the bacteria without exposing them to toxic peroxide concentrations or allowing too rapid decomposition of the peroxide to gaseous oxygen which can plug the aquifer. Until recently the conventional approach of pump-and-treat was the common and preferred method at most sites. The more up-to-date engineers, however, seriously evaluate new techniques, such as in-situ biodegradation.

Conclusions:

There are many other examples of important problems facing the ground water engineer, but the ones discussed here are surely representative. If one could sum up the most important guidance to newly initiated ground water engineers, I believe it should be to realize the importance of the initial analysis or assessment of the contamination, its extent, its severity, and then an objective estimate of the real risk that the contamination presents. This will set the stage for providing the client the most cost effective solution.

Control of Groundwater Contamination: Case Studies

James W. Mercer*, Charles R. Faust*, Anthony D. Truschel*
and Robert M. Cohen*

Abstract

     Engineered solutions to groundwater contamination problems are
being proposed at many hazardous waste sites, yet the effectiveness
and long-term reliability of many of these proposed solutions are
unknown. Only recently have sufficient data on existing engineered
solutions been collected to enable preliminary evaluation; however,
some of this information is difficult to obtain due to the
litigation surrounding most sites. Case studies are an important
method of identifying problem areas in controlling contamination.
Knowledge from one site can provide information about corrections
and improvements to engineered solutions at other hazardous waste
sites. Three case studies are discussed in this paper, including
Love Canal near Niagara Falls, New York, Rocky Mountain Arsenal near
Denver, Colorado, and Lipari landfill near Pitman, New Jersey.
Among these three sites, a variety of remediations have been
utilized, including recharge and discharge wells, low-permeability
walls, drains, and various types of covers. The effectiveness of
each method is evaluated using groundwater monitoring data. Results
of the evaluation reveal a wide range in the level of success of the
various remediations, thus indicating the importance of monitoring
following remediation.

Key Words: Groundwater, contamination, remediation, Love Canal,
            Rocky Mountain Arsenal, Lipari.

INTRODUCTION

     The clean up of hazardous waste sites has received much
attention since the enactment of Superfund (CERCLA) in 1980. While
USEPA estimates that about 2,000 sites will reach the National
Priorities List, US Office of Technology Assessment (OTA, 1985)
estimates that at least 10,000 sites may require clean up by
Superfund. Many of these will involve groundwater clean up; as
indicated by EPA (1984), of the 395 uncontrolled hazardous waste
facilities studied, 68 percent involve groundwater contamination.
Remedial actions at most of these sites will involve some type of
hydraulic control, ranging from reducing recharge to groundwater
extraction.

     When groundwater is a concern at hazardous waste sites,
characterization and remediation of the site are generally more
time-consuming and expensive for the following reasons: to reduce
uncertainties associated with the subsurface, wells must be drilled;
*GeoTrans, Inc., Herndon, VA

the time required to clean up groundwater is lengthy, primarily due
to slow groundwater velocities. Because of the expense and the long
time frames, and because of the variety of available remediation and
treatment technologies (OTA, 1984), selection of the appropriate
remedial action is both difficult and important.

Mercer et al. (1985) discuss the use of groundwater models to
aid in this selection process. They point out that only long-term
monitoring will verify whether engineering design criteria are met
and if system behavior is sufficiently understood. Although it is
too early for long-term monitoring data at most hazardous waste
sites, several sites exist where remedial actions were implemented
several years ago and subsequent monitoring data are available.

This paper will review existing monitoring data at three sites
and evaluate the remedial actions in terms of cost and effectiveness
of cleaning up groundwater. The Love Canal site in Niagara Falls,
New York has been extensively studied, and involves the use of a
synthetic cover and a French drain. The Rocky Mountain Arsenal near
Denver, Colorado is another site that has been extensively studied
(e.g., Thompson et al., 1986). The Arsenal's North Boundary
Containment System, considered in this paper, consists of a slurry
wall in combination with groundwater extraction and injection wells.
The final case history concerns the Lipari site near Pitman, New
Jersey, and currently involves an encircling slurry wall and
synthetic cap. Each of these sites is discussed below. It is hoped
that presentation of case histories in this fashion will provide
guidance in remedial action selection at future sites.

CASE HISTORIES

LOVE CANAL

Brief History. Located near Niagara Falls, New York, Love
Canal was excavated in the 1890s to enable generation of
hydroelectric power for development of a "model" manufacturing city.
Excavation of the proposed 6 to 7 mile canal had barely begun, with
approximately 3000 ft dug just north of the upper Niagara River,
when the project failed in 1896. The property was used for
recreation by a growing residential community when the Hooker
Electrochemical Company began dumping chemical wastes at the site in
1942. Prior to waste disposal at the site, Love Canal was
approximately 3000 ft long, 40 to 100 ft wide, and 8 to 15 ft deep.
Hooker purchased the property in 1947 and then sold it to the
Niagara Falls Board of Education for $1 in 1953. The Board of
Education wanted the property adjacent to the canal for construction
of an elementary school. Between 1942 and 1953, Hooker buried
approximately 22,000 tons of chemicals (primarily organic compounds
including chlorinated benzenes, toluene, lindane, and
trichlorophenol) at the site, according to company estimates.

After sale of the property to the Board of Education,
development of the residential area around Love Canal continued into
the 1970s. Although a variety of problems related to the landfill
periodically surfaced subsequent to site closure in about 1953, it

was not until the mid-1970s that the landfill attracted public attention. The development of a very high water table, due in part to heavy precipitation, exacerbated a number of problems: (1) subsidence of the landfill surface and exposure of drums, (2) ponding of contaminated surface water in some backyards adjacent to Love Canal, (3) the presence of unpleasant chemical odors which were cited by residents as a cause of discomfort and illness, (4) the migration of contaminated groundwater and non-aqueous phase liquids (NAPL) into basements adjacent to the landfill, and (5) the migration of chemicals into and through the local sewer system. As a result of these and related problems, several Health Emergencies were ordered by the New York State Health Commissioner in 1978, and a State-of-Emergency was declared twice by President Carter, once in 1978 and again in 1980.

Remedy. Numerous investigations have been conducted since 1977 to assess the extent of environmental contamination in the Love Canal area and to recommend remedial actions. Remediation at the site has proceeded in phases. The main objectives of remediation have been to contain chemical migration, clean up contaminated areas, and limit chemical exposure. Implemented remedies have included site evacuation, demolition of homes and a school, installation of clayey soil and synthetic covers, construction of a French drain leachate collection system with a carbon adsorption treatment plant, severance and plugging of utility lines, construction of cutoff walls at selected locations, installation of fencing, and the cleaning of sewer lines. Future remedial work is scheduled to include clean up of local creeks and continued operation and maintenance of the site containment system and long-term monitoring program.

Phase 1 of site remediation occurred in the south section of Love Canal between October 1978 and February 1979. A leachate collection system and clayey soil cover were constructed and leachate was treated using temporary facilities. The leachate collection system consisted of barrier drains that were installed parallel to the southern portion of Love Canal and midway between the landfill edge and the formerly adjacent homes. Vitrified-clay pipe with an 8-in diameter was placed in trenches that were dug 12-15 ft deep and 4 ft wide. Approximately 2-3 ft of crushed stone were placed around the tile pipe and the trench was backfilled with sand to ground level. Leachate entering the drains flows by gravity (0.5% grade) into concrete wet wells. From the wet wells, leachate is pumped to subsurface holding tanks and then to the onsite treatment plant. In January 1979, eight lateral French drains were constructed from the main drains to the landfill to expedite dewatering of the site prior to the installation of the clayey soil cover. The cover consisted of approximately 3 ft of clayey soil at the apex with a gradual tapering at the sides and about 1 ft of soil over the tile drains. Treated leachate is discharged to the sanitary sewer system.

Phase 2 of the remedial program began in May 1979 and was completed in December 1979. Changes in remedial design during Phase 2 included the following: (1) the drains were located as deep as 18

ft rather than 12-15 ft; (2) 6-in rather than 8-in diameter
vitrified-clay pipe was used; (3) a berm was built to prevent
surface runoff of contaminated water from the site; and (4) a steel
wire mesh was placed around each pipe joint for reinforcement. As
in the southern section, numerous French drain laterals were
installed between the main drain lines and the landfill to speed
site dewatering. The discovery of large amounts of leachate and
porous fill in a filled drainageway (swale) on the northwest side of
the landfill prompted placement of a clay cutoff wall, 11 ft deep
and 3 ft wide, on the west side of the buried swale. Several foot
drains and pipes uncovered along the the trenchline were filled with
cement mortar. During this period, an examination of the barrier
drain in the southern section revealed numerous pipe separations and
infilling of stone. As a result, 67 percent of the southern tile
drain was repaired between July and October of 1979. Finally, the
permanent, activated carbon leachate treatment plant was built and
put into operation in December 1979.

By May 1982, approximately 570 families had been relocated from
the Emergency Declaration Area (EDA). Other remedial actions taken
included: (1) the plugging of the Wheatfield Avenue sanitary sewer
in November 1980 following determination that Love Canal chemicals
were migrating away from the landfill through this pathway, and (2)
final work on the clayey soil cover over the landfill which was
completed in 1981.

A storage facility was constructed between April and August of
1982 for sludge that is collected by the barrier drain system but
cannot be treated at the carbon adsorption treatment plant.
Recently, plans have been made to thermally destroy sludge that has
accumulated at the treatment plant. Between June to August 1982,
227 homes surrounding Love Canal were demolished. Building debris,
concrete walls, patios, etc. were placed in basements and covered
with clay. All utilities and sewer lines to the homes were
terminated and sealed prior to demolition.

New remedial work to improve the site containment system began
in September 1982 and was completed in December 1984. New phase 1
work occurred between September 1982 and March 1983 and included:
(1) cutoff and plugging of existing onsite utility lines; (2)
construction of a new sewer line to the treatment plant; and (3)
construction of offsite drainage facilities. New phase 2 work
between April 1983 and December 1983 included: (1) stripping of
topsoil and recompaction of the existing clayey soil cover; (2)
inspection, cleaning, and repair of the existing barrier drains; (3)
placement of imported earthfill; (4) placement of concrete cutoff
walls at 7 street crossings; (5) demolition of imported earthfill;
and (6) demolition of the 99th Street School. In Phase 3, from May
to December 1984, the following work was completed: (1) placement
of imported earthfill and recompaction of the existing clay cap; (2)
installation of a HDPE synthetic membrane cover; (3) installation of
drain tiles on top of the HDPE cover to divert surface runoff; (4)
abandonment and/or improvement of roadways; and, (5) extension of 2
of the 4 pump stations serving the barrier drain system. The
concrete cutoff wall surrounding the site was not constructed as

planned because chemicals had been discovered along portions of the wall alignment and additional analysis indicated that the benefits of the wall would be relatively minor.

A remedial investigation to determine the extent of chemical migration in area sewers and creeks was conducted in 1983. Sewer clean up began in April 1986. Truck mounted hydraulic vacuum systems were used to remove sediment from the sewer lines. The sewer sediments will be stored in drums in the Love Canal containment area. As of September 1986, clean up of the local creeks and the upper Niagara River has not begun.

Cost. As of December 1983, the United States had expended approximately $45,000,000 in studies and response actions at Love Canal (Cohen, 1983). The State of New York and City of Niagara Falls also have committed vast resources to the problem. Remedial project costs, particularly during the early phases of remediation, have been difficult to predict (Glaubinger et al., 1979). Over the long term, additional costs will be incurred in the operation and maintenance of the site containment system and site monitoring.

Effectiveness. Construction of a leachate collection and treatment system and placement of a clayey soil cover over the Love Canal landfill between 1978 and 1981 provided substantial environmental benefits to the community. Installation and maintenance of the clay cover has been effective in the following ways: (1) preventing the surface runoff of chemicals from Love Canal; (2) preventing human contact with chemicals at the surface of the landfill; (3) reducing the rate of recharge to the landfill; and (4) improving the appearance of the site. Construction of a barrier drain leachate collection system has improved drainage at the landfill, thereby preventing the water table from overflowing the Love Canal "bathtub"; provided for collection of hazardous chemicals in the subsurface for treatment; and prevented and/or reduced the previously uncontrolled migration of hazardous chemicals from Love Canal through the overburden sediments.

Benefits were also realized from the improvements made to the site containment system between 1982 and 1984. Severance and plugging of utility lines leaving the site has eliminated preferential pathways for chemical migration that were active prior to remediation. Hydraulic analysis (Cohen and Mercer, 1984) of a proposed concrete cutoff wall and extended synthetic membrane cover in 1983 indicated that the synthetic cover, in particular, would be effective in reducing both infiltration to the landfill and groundwater inflow to the drain system, and in extending the influence of the drain system. Monitoring data has shown that flow to the leachate collection system has been substantially reduced since cover installation, indicative of reduced infiltration. Determination of the zone-of-influence of the drain system has been difficult, however, due to the complex nature of the overburden and the flatness of the water table at the site. Recent installation of additional nested piezometers in transects perpendicular to the barrier drains should improve determination of the capture zone of the drains.

Adequate monitoring is required to assess chemical migration from hazardous waste sites. Chemical migration at Love Canal went largely unnoticed until chemicals and odors seeped into basement sumps and overflowed the landfill, collecting in surficial pools. Similarly, monitoring is required for determining the effectiveness of remedial actions.

The goals of the long-term monitoring program at Love Canal are threefold: (1) to provide an early warning of renewed chemical migration from the site in the future; (2) to provide data on the effects and effectiveness of the site containment system; and (3) to provide additional data which can be used to assess the habitability of the EDA (E.C. Jordan, 1985). A ring of monitoring wells, with a spacing of approximately 300 ft, has been installed around the site to enable detection of renewed chemical migration through the overburden. Series of nested piezometers (with individual piezometers completed to the different overburden units at each location) have been constructed in transects perpendicular to the barrier drains to monitor the effectiveness of the site containment system. Additional bedrock wells also have been drilled to investigate the presence of Love Canal chemicals and to enhance the bedrock aquifer monitoring capability. Finally, monitoring of surface water and sewer stations has been proposed.

In summary, remedial actions at Love Canal have been multifaceted and complex. They have occurred over a long period of time, from 1978 to present. Monitoring has continued throughout this process, resulting in improved remediation, e.g., clay cover was replaced and extended by a synthetic cap. Continued long-term monitoring is an integral part of the remedial program. Costs are considerable and have been difficult to estimate.

ROCKY MOUNTAIN ARSENAL

Brief History. The Rocky Mountain Arsenal (RMA) occupies approximately 17,000 acres in Adams County, Colorado. The RMA was established in 1942 to manufacture and process chemical warfare products. In 1946, portions of the RMA were leased to private industry for the manufacture of pesticides and herbicides. From 1943 to 1956 chemical wastes were disposed in unlined basins. Evidence of off-post migration of groundwater containing contaminants matching those disposed in the unlined disposal basins led to the construction of an asphalt-lined evaporation basin. It was subsequently concluded that the lined basin failed (Konikow and Thompson, 1984).

Although contamination problems and remediation have occurred at various locations on the RMA, only the north boundary is considered in this paper. In the mid 1970's, RMA specific contaminants diisopropylmethylphosphonate (DIMP) and dicyclopentadiene (DCPD) were found in groundwater and surface water north of the RMA boundary. In response to the observed off-post contamination and the Colorado Department of Health's (CDH) cease and desist orders, a program was established that included groundwater monitoring to determine a means to intercept

contaminants flowing across the north boundary of the RMA. As a
result of continued monitoring, additional contaminants were
identified, including Nemagon or dibromochloropropane (DBCP).

Remedy. The remedial concept selected involved interception of
groundwater a short distance south of the northern RMA boundary,
treatment of the water to remove contaminants, and injection of the
treated water at the boundary (Konikow and Thompson, 1984). A pilot
groundwater contamination control system was initiated in 1977 and
completed in 1979. The purpose of the pilot system was to evaluate
the feasibility of the selected remedial concept. The North
Boundary Pilot System (NBPS) consisted of six alluvial withdrawal
wells and 12 alluvial injection wells, with a 1500-ft long, 3-ft
wide slurry wall in between. The primary purpose of the slurry wall
was to reduce the potential for excessive recycling of water from
the injection wells back to the withdrawal wells. The wall was made
with a mixture of soil and bentonite clay anchored into
approximately 2 ft of bedrock. The dewatering wells were
approximately 225 ft apart, were placed in a straight line parallel
and upgradient of the wall, and were 8-in diameter wells placed
within 30-in diameter gravel packed holes. Each well was screened
throughout the saturated portion of the alluvial aquifer. The
injection wells were approximately 100 ft apart, were placed in a
straight line parallel and downgradient of the wall, and were 18-in
diameter wells placed within 36-in diameter gravel-packed holes.

It was concluded that the interception and treatment of
alluvial groundwater contaminated with DIMP and DCPD was feasible
and therefore expansion of the NBPS was proposed. In 1979, the CDH
requested that the expanded NBPS intercept and treat all groundwater
contaminated with more than 0.2 ug/L DBCP, in addition to
groundwater contaminated with DIMP and DCPD.

Construction of the expanded containment system was initiated
in 1979 and completed by January 1982. The slurry wall was expanded
3840 ft to the east and 1400 ft to the west. The expansion of the
containment system was based on historical DIMP, DCPD, and DBCP
plume migration pathways within the alluvial aquifer. The total
depth of the expanded slurry wall was also increased to intercept
shallow sand/sandstone deposits in the bedrock; the slurry wall
ranges in depth from 20 ft at the NBPS where it does not intercept
any bedrock sand/sandstone deposits, to approximately 40 ft along
the eastern extension. The total number of activated carbon columns
was increased from two to three. A new treatment system, capable of
treating 600 gal/min replaced the pilot plant (Thompson et al.,
1986). The number of withdrawal wells was increased to a total of
54 and the number of injection wells was increased to a total of 38.

Cost. The cost of the barrier and the wells as constructed in
1978 was $450,000 (Konikow and Thompson, 1984). The facility for
housing the treatment system cost approximately $40,000. The
treatment equipment was leased with an associated initial cost of
approximately $100,000 and a yearly fee ranging from $135,000 to
$150,000 (Konikow and Thompson, 1984). The new expanded system
began operation in 1982 at a cost of $6 million (Thompson et al.,

128   CONTAMINATED GROUND WATER

1986). Each year the Army spends an additional $1.4 million to
operate the facility. The original estimate for the expanded system
construction, operation, and maintenance over a 25-year period was
approximately $6.7 million (D'Appolonia, 1979).

Effectiveness. Numerous problems with the operation of the
containment system have been identified. These operational problems
include freezing of check valves, balancing valves and flow meters
at the well heads, and electrical and mechanical control systems
problems, especially as a result of damage from lightening (Thompson
et al., 1986). A major problem is the system's inability to inject
the same volume of water removed by the withdrawal wells. One
reason for this is that the design and number of injection wells
installed was marginal with respect to the volume of groundwater
extracted. The second reason is that the efficiency of the
injection wells has been reduced due to suspended sediments (small
particles of carbon which were escaping from the adsorbers) in the
injected water plugging the formation. This potential problem was
pointed out in the feasibility study of the North Boundary
Containment System (D'Appolonia, 1979).

Because all treated water could not be injected, some treated
water was discharged to nearby surface water. This is a deviation
from the proposed design and operation of the system. The system
was designed such that the spatial distribution of injection wells
would balance the distribution of withdrawal wells, thereby
minimizing the amount of change to the precontainment flow field.
By discharging treated water to the surface, the groundwater flow
field is altered. In addition, discharging at the surface has
increased the potential evaporation rate, thereby reducing the
volume of water that is ultimately returned to the groundwater
system.

Alluvial aquifer monitoring wells selected for use as off-post
detection wells have shown concentrations of DIMP and other RMA
specific contaminants (the Denver Post, July 10, 1986). The
presence of DIMP in these wells, along with other RMA contaminants,
indicate either: (1) continued migration of RMA contaminants either
through, under or around the containment system; (2) the
ineffectiveness of the containment system to adequately intercept
contaminated groundwater; (3) the slow release of DIMP from mineral
surfaces (desorption) in the contaminated aquifer; or (4)
groundwater contamination prior to the construction of the
containment system.

DIMP also has been detected in one bedrock aquifer monitoring
well included in the off-post monitoring program (the Denver Post,
July 10, 1986). Its presence may be from groundwater contamination
prior to the construction of the containment system.

It has been assumed historically that contamination of the
bedrock aquifer is constrained to the shallow sand/sandstone
deposits. It has also been assumed that any deep contamination
on-post would eventually move up-dip (parallel to bedding) and back
into the alluvial aquifer before migrating off-post (May, 1982).

Limited data exist that indicate the lack of continuity of sand/sandstone deposits in the northern direction near the lined basin. Thus, groundwater flow may not be constrained to follow the dip of the sand/sandstone deposits, allowing contaminants to move deeper through these units and beyond the influence of the containment system.

In summary, limited data indicate some problems with the operational effectiveness of the North Boundary Containment System. Specifically, the present system configuration does not enable the injection of the same volume of water removed by the withdrawal wells. Contamination of the alluvial aquifer and under bedrock aquifer has been found, indicating that the containment system is perhaps failing to prevent continued off-post migration of RMA contaminants.

It is important to note that substantial amounts of data have been generated recently that yield additional insight to the effectiveness of the North Boundary Containment System. At this time, however, the data can not be presented due to current litigation involving the Rocky Mountain Arsenal clean up and natural resources damages.

LIPARI LANDFILL

Brief History. The Lipari landfill is an inactive, 16-acre landfill site, 6 acres of which were formerly used for the disposal of industrial and municipal waste. The site was purchased in 1958 for use as a sand and gravel pit. Liquid wastes were emptied into the landfill from 1958 to 1969, and solid wastes were disposed until May 1971 when the site was closed by the New Jersey Solid Waste Authority. The wastes reported to have been disposed included solvents, paint thinners, formaldehyde, paints, phenol and amine wastes, dust collector residues, resins, and ester press cakes (Camp Dresser and McKee, 1985). The site was closed as a result of a cease and desist order directed at pollution of surface water as a result of off-site leachate migration.

Remedy. In August 1982, the USEPA issued a Record of Decision (ROD I) which outlined a two-phase approach to remediation of the Lipari landfill and selected the Phase I approach. This approach involved the containment of the landfill using a soil-bentonite slurry wall and synthetic cap. Remedial construction activities began on September 7, 1983 with the installation of the 30-in wide slurry wall completely encircling 15.3 acres of the site. The slurry wall was keyed into an underlying clay layer that ranged in depth from about 15 to 55 ft. In December 1983, field operations were halted due to inclement weather before the synthetic membrane cap could be completed. The following spring, water levels inside the containment system rose, due to snow melt and heavy rainfall, to levels which prohibited completion of the cap. The cap could not be completed until after water levels were lowered in September and October 1984 using extraction wells. The completed synthetic cap consists of 40-mil thick, high density polyethylene, and includes a passive gas-venting system and surface water drainage system.

The Phase II objective, as defined by the EPA in ROD I, is to improve the reliability of the containment system. The Phase II program selected by the EPA in its second Record of Decision (ROD II) on the Lipari landfill, dated September 30, 1985, entails repetitive flushing of the entire 16-acre site over at least a 15-yr period. To date, this new effort has not been implemented.

Cost. The Phase I costs at the Lipari landfill are estimated to be $3,956,492.62 (Letter to Bradford F. Whitman from Lawrence R. Maddock, November 21, 1985). This includes both studies and the actual remedial program, with the latter estimated at $2,550,035.07. Cost estimates prior to construction could not be obtained.

Effectiveness. The slurry wall was designed to have a hydraulic conductivity of 10^{-7} cm/s. Records indicated that sidewall cave-ins and sediment events occurred during construction. Remedial efforts were taken to correct these problems, but no direct evidence was obtained to insure that cave-in or sediment materials were removed (Camp Dresser and McKee, 1985). However, based on a review and analysis of the construction logs, QA/QC records of the contractor, and permeability testing, Camp Dresser and McKee (1985) estimated that the bulk hydraulic conductivity of the wall was 10^{-7} cm/sec or less. Water-level data taken after installation of the wall and cap suggest that the bulk hydraulic conductivity is substantially greater. These data show a consistent pattern of high water levels in the southwest portion of the containment and lower levels in the northeast portion; thus indicating substantial groundwater flow from the southwest to the northeast. Using this information, it was shown (Betz Converse Murdoch, 1986) that the range for the wall hydraulic conductivity is 1.0 to 5.0 x 10^{-6} cm/s, which is ten to fifty times more permeable than that originally estimated. More recently, Camp Dresser and McKee (1986) has reevaluated all existing information including recent water-level data and has concluded that the bulk hydraulic conductivity of the wall is between 4 x 10^{-7} and 4 x 10^{-6} cm/s. The significance of the revised estimates is that much more leakage from the containment wall (4 - 50 times as much) is occurring than originally assumed.

Existing water-level trends within the containment suggest that the synthetic cover has been effective. The data show a relatively steady decline in water levels over the past year and a half. The containment water levels are declining toward a new equilibrium with water levels on the outside of the containment and in the lower aquifer. The water-level trends on the inside do not show fluctuations that would occur in response to recharge conditions.

Original estimates of the wall and cap effectiveness were used by Camp Dresser and McKee (1985) in an analysis that was the basis for selecting a Phase II remedy. An important factor in this analysis was the relative importance of leakage from the walls versus leakage through the clay layer below the containment.

Because the wall is apparently much more permeable than previously assumed, the negative impact of the selected Phase II remedy during periods of high water levels needs reevaluation. A more recent analysis by Camp Dresser and McKee (1986) has considered revised estimates of hydraulic conductivity. This analysis led to the conclusion that an offsite collection system above the clay layer would be required if the flushing remedy is implemented.

In summary, construction difficulties during implementation of the Phase I remediation resulted in the contained area becoming saturated and delayed completion of the cap. Evaluation of the wall hydraulic conductivity using water-level monitoring data indicated a value that is four to fifty times greater than that originally estimated. The significance of corresponding reestimates of leakage through the containment walls needs to be further considered in the selection of Phase II remedies for the site.

CONCLUSIONS

Three hazardous waste sites involving groundwater contamination have been reviewed in an effort to summarize effectiveness and costs of remedial actions. Several conclusions are made based on this review:

(1)   Hazardous waste sites involving groundwater contamination generally require more time and effort to characterize and remediate than sites not involving groundwater contamination.

(2)   Good pre-remedial site characterization is critical to both selection and implementation of remediation. Because of seasonal changes in groundwater, a minimum of one year should be devoted to monitoring and characterization before a remedy is selected. As the site complexity increases, this time will increase proportionately.

(3)   In order to minimize costs, both site characterization and remediation should be performed in phases, such that later phases may be modified based on knowledge gained from earlier phases.

(4)   As the scale of the observation increases, properties, such as permeability, tend to increase because more heterogeneities are encountered. Therefore, remediations based on core-scale observations, may underestimate groundwater flow rates.

(5)   Site characterization and remediation tend to be costly at sites involving groundwater contamination, with clean up costs difficult to estimate accurately.

(6)   Monitoring is critical for both site characterization and remediation. Long-term monitoring should be an integral part of any remedial action plan. In addition, it is important to monitor before, during and after remediation in

order to evaluate effectiveness. Groundwater elevation data, which is relatively inexpensive to obtain, can be particularly useful in the evaluation of remedial effectiveness.

(7) The effectiveness of various remediations varies from site to site, and depends in large part on the site characterization and analysis. Of particular importance at hazardous waste sites is the lack of good bedrock characterization prior to remediation. Apparent containment can be lost because of unexpected flow through the bedrock (in addition to some cases presented in this paper, for example, see Ozbilgin and Powers, 1984, concerning the site in Nashua, New Hampshire).

REFERENCES

Betz Converse Murdoch, Inc., "Phase II Remedial Action Recommendations," Lipari Landfill, Feb., 1986.

Camp Dresser and McKee Inc., Final Draft Report, "Onsite Feasibility Study for Lipari Landfill," EPA Contract No. 68-01-6939, August, 1985.

Camp Dresser and McKee, Inc., Final Report, "Technical Response To Phase II Remedial Action Recommendations For Lipari Landfill - Submitted by Rohm & Haas Company (February, 1986)", EPA Contract No. 68-01-6939, July 1986.

Cohen, R.M. and J.W. Mercer, "Evaluation of a Proposed Synthetic Cap and Concrete Cut-Off Wall at Love Canal Using a Cross-Sectional Model," Fourth National Symposium and Exposition on Aquifer Restoration and Ground-Water Monitoring, Columbus, OH, May 1984.

Cohen, A., "Second Amended Complaint of the United States of America," US, NY, and UDC-Love Canal Inc. v. OCC, Civil No. 79-990, 1983.

D'Appolonia Consulting Engineers, Inc., Conceptual Design of the North Boundary Containment System, Rocky Mountain Arsenal, Denver, Colorado, 1979, 37 pp.

D'Appolonia, Waste Management Service Letter Report, "Long-Term Permeability Testing Slurry Trench Backfill Lipari Landfill," Pitman, New Jersey, October 13, 1983.

E.C. Jordan Co., "Love Canal Remedial Project Long-Term Monitoring Program Design," Report to New York Department of Environmental Conservation, 1985.

EPA, "Summary Report: Remedial Response at Hazardous Waste Sites, U.S. Environmental Protection Agency," EPA-540/2-84-002a, 1984.

Glaubinger, R.S., P.M. Kohn and R. Ramirez, Love Canal Aftermath, Chemical Engineering, 1979, pp. 86-92.

Konikow, L.F., and D.W. Thompson, "Groundwater Contamination and Aquifer Reclamation at the Rocky Mountain Arsenal, Colorado," Groundwater Contamination, National Academy Press, Washington, DC, 1984, pp. 93-103.

May, J.H., "Regional Groundwater Study of Rock Mountain Arsenal, Denver, Colorado Report I: Hydrogeological Definition," 1982, 78 pp.

Mercer, J.W., C.R. Faust, R.M. Cohen, P.F. Andersen, and P.S. Huyakorn, "Remedial Action Assessment for Hazardous Waste Sites via Numerical Simulation," Waste Management & Research, vol. 3, 1985, pp. 377-387.

OTA, Protecting the Nation's Groundwater From Contamination, U.S. Office of Technology Assessment, OTA-O-233, Washington, DC, 1984.

OTA, Superfund Strategy, U.S. Office of Technology Assessment, OTA-ITE-252, Washington, DC, 1985, 282 pp.

Ozbilgin, M.M. and M.A. Powers, "Hydrodynamic Isolation in Hazardous Waste Containment, Fourth National Symposium and Exposition on Aquifer Restoration and Ground-Water Monitoring, Columbus, OH, USA, May 1984.

Thompson, D.W., E.W. Berry, and B.L. Anderson, Clean-up in the Rockies, Civil Engineering, American Society of Civil Engineers, February, 1986, pp. 40-42.

BEST MANAGEMENT PRACTICES FOR POINT AND NONPOINT
SOURCES OF GROUNDWATER CONTAMINATION

Martin Jaffe[*]

ABSTRACT

A number of states and communities have adopted programs to address groundwater contamination from both point and nonpoint sources.
The regulation of point sources has generally been given the highest
management priority, consistent with federal environmental protection statutes (such as the 1984 RCRA amendments) requiring that these
contamination threats be addressed by states as a condition of receiving federal funding and program support. Management strategies
have included regulations to guide or limit development in aquifer
recharge areas and in the capture zones of public wellfields, engineering requirements to minimize infiltration into and exfiltration
from discrete sources, and administrative innovations which expand
or modify agency responsibilities with respect to identified sources
posing groundwater threats.

The management of nonpoint sources is more problematic. The
wide geographic area of application, the relatively low toxicity of
many of these contaminants, and the scope of their risks (typically
to on-site domestic water supply wells and not to public wells) often
make them a lower management priority in many communities. However,
a number of communities and state agencies are beginning to examine
the incremental impacts of nonpoint sources on groundwater resources
and have started to fashion management programs to address these
risks. As nonpoint sources achieve a higher management priority,
they are likely to become subjected to new regulatory programs and
requirements. Currently, most nonpoint sources are managed through
voluntary compliance and educational programs. A key factor in the
success of new regulatory efforts will be the administrative mechanisms that can be created to execute, monitor, and enforce the regulatory requirements.

THE CONTROL OF POINT SOURCES

State and local governments have long addressed many discrete
sources of groundwater contamination as part of their general programs for environmental management and protection. Solid and
hazardous waste landfills, for example, have been regulated by many
states as part of their efforts to comply with federal requirements
set forth in the Solid Waste Management Act, and the amendments to
the Act created by the 1976 and 1984 Resource Conservation and Re-

* Associate Professor, School of Urban Planning & Policy, University of Illinois at Chicago, P.O. Box 4348, Chicago, IL
60680.

covery Act. Wastewater discharges have been addressed by state and
local plans and programs to establish wastewater treatment facilities
and sewer extensions under the Clean Water Act and its amendments.
In addition, various discrete sources of contamination--such as underground storage tanks and industrial waste impoundments--currently
are being brought under regulatory control as a result of recent federal legislation (the 1984 amendments to the Resource Conservation
and Recovery Act) addressing these pollution sources.   The federal
Superfund law has also directed attention towards the identification
and management of abandoned hazardous waste sites posing discrete
pollution threats to ground and surface water supplies and to environmental resources.

     To a large extent, these state and local programs arose in reac-
tion to federal program initiatives, particularly funding opportunities tied to compliance with federal program requirements. As a
result, two major features characterise the programs that have been
developed from these federal guidelines: first, the reactive status
of state and substate efforts have resulted in program uniformity
with respect to the specific sources addressed by the legislation
and funding programs, and, second, sources have tended to be addressed in response to funding priorities and not threat priorities
facing a state or substate region.   In other words, since federal
program requirements determined funding eligibility, most programs
have achieved a high degree of conformity with the federal regulations resulting in a fairly uniform management response across the
nation.   Where state or local policies diverged from such programmatic requirements, the political entities have tended not to participate in the management program by not achieving primacy (leaving
uniform management to the U.S. EPA, which retained regulatory responsibility over the sources).   One characteristic feature of source
control, then, is that states and localities all tend to regulate
priority contamination sources in much the same way; there tends to
be fairly uniform "best management practices" depending on the program requirements set forth under the federal legislation addressing
a particular pollution threat.

     This national uniformity has several implications, including the
second distiguishing feature of many source control programs--their
tendency to be put into place without regard to localized contamination risks or their relationship to local groundwater resources.   In
some cases, as with wastewater facilities planning and construction,
sewer lines and treatment plants were established irrespective of
overriding environmental considerations and merely because federal
subsidies were available to encourage local growth and development
beyond that which could be sustained in the absense of such incentives.   Competition for taxable growth gave communities an incentive to build new sewer facilities or to establish new landfills
meeting federal program requirements; this same competition raised
significant growth management issues, including extraterritorial
utilities expansion linked to future annexation and the encourage-
ment of industrial and commercial development in locations that were
particularly susceptible to groundwater contamination incidents by
materials spills and improper disposal and storage practices.

These trends resulted in the third distinguishing characteristic of pollution source controls, their fragmentation. Federal legislative initiatives, and the programs that they encouraged, for the most part address discrete threats from individual generic sources, but do not address synergistic threats or threats arising from unregulated sources. Efforts to comprehensively manage a variety of sources have arisen with respect to the U.S. EPA's consolidated permitting program, under program requirements mandating the consideration of secondary impacts (such as the integration of air quality considerations from sources served by wastewater treatment facility expansions funded under the Clean Water Act), and under comprehensive planning programs (such as areawide water quality management planning that addressed source controls directed towards surface water protection efforts). Although a step in the right direction, these initiatives were not developed to comprehensively manage groundwater threats posed by the generic sources. The planning requirements in the Sole Source Aquifer Demonstration Program and in the Wellhead Protection Area requirements of the recent Safe Drinking Water Act amendments may go far in promoting a more comprehensive management approach with respect to demonstration projects in designated sole source aquifers and in the designated wellfield protection areas of public water supply wells.

Best Management Practices for Generic Sources

Many contaminant sources already are regulated under federal legislation, and this has established the range of actions often employed by state and local agencies. Typically two approaches predominate--identification and evaluation of individual sources followed by engineering standards to minimize the potential for infiltration into the source and exfiltration from the source. Source surveys are used to identify hazardous waste facilities, solid waste facilities, industrial waste impoundments and lagoons, and underground storage tanks. In many cases, these inventories arose from earlier 208 areawide water quality management planning.

Engineering standards often are fairly straightforward. Infiltration of precipitation into a source and exfiltration of materials from a source are addressed by the use of liners, vaulting, or the employment of impermeable materials around the source. Again, these standards often appear in federal regulations addressing generic sources (i.e., as under RCRA). Wastewater effluent discharges also are addressed by minimizing exfiltration from sewers in sensitive areas (such as those with highly permeable soils overlying shallow unconfined aquifers), and by standards governing the permeability of soils surrounding leaching fields of on-site systems (by establishing maximum and minimum percolation rates for system approval and installation).

These same principles have been applied by local governments to activities and sources that are not directly addressed by relevant federal legislation. Besides on-site wastewater disposal facilities, solid waste disposal facilities probably constitute the most common source regulated under state standards, but the same engineering ap-

proaches are employed--liners and impermeable caps--as with those
sources regulated under federal law. Some local governments--such
as Spokane County, Washington--have broadened the range of potential
sources to include industries and businesses using and storing toxic
materials on-site. These also have been inventoried, with "best
management practices" again relying on the use of impermeable surfaces
to minimize infiltration and exfiltration from spills. Similar approaches have also been used for animal feedlots, above-ground storage tanks, and loading areas in warehouses and industries.

Recent Trends in Point Source Controls

    A number of local governments have shifted away from engineering
measures and instead have employed land use control approaches to
regulate point sources. These newer approaches have relied on locational restrictions and on density requirements to minimize discharge
threats from generic sources. Locational restrictions are best ex-
emplified by Dade County, Florida's wellfield protection program,
where the intensity of activities with the potential to pose groundwater threats are reduced the less the hydraulic travel time to the
wellhead. Reducing housing densities, for dwellings relying on on-
site wastewater disposal systems, in areas of high permeability or
low depth to groundwater is another land use control measure that has
been employed on Cape Cod, Massachusetts, and on Long Island, New
York. In some communities, such as in Crystal Lake, Illinois, per-
missible densities may be decreased as much as one dwelling unit per
forty acres in extremely sensitive areas.

    A second trend is to favor preconstruction review and hydrogeo-
logic assessments as a condition of development approval. Communities in Chester County, Pennsylvania, for example, require the sub-
mission of a geological report when development is proposed in an
area overlying a permeable carbonate aquifer transecting the county.
This technique shifts the data collection and evaluation burden onto
the person proposing to establish a potential contamination source,
conserving scarce local resources and staff. Similar information
disclosure requirements have been recommended in carbonate and karst
areas of southeastern Minnesota by local environmental groups, and
in areas of Ohio where mining, oil drilling, and other extractive
industries pose groundwater threats through aquifer interconnection
and brine disposal.

    A third approach is merely to ban certain potential sources in
sensitive areas. These approaches may be carried out by a variety
of regulatory vehicles besides land use controls, including health
codes and general police power ordinances in addition to eminent
domain authority. The approach used to acquire property in wellhead
capture zones in Massachusetts and in Schenectady County, New York,
for example, relies on eminent domain, while the ban on septic system
additives by Suffolk County, New York, is in the form of a general
police power ordinance. Mandatory separation distances between septic system leach fields and on-site drinking water supply wells is
a traditional approach found in many state and local health codes;
unfortunately, in many cases such restrictions serve only to guide

138   CONTAMINATED GROUND WATER

the emplacement of new wells, but not subsequent septic system locations. Buffer areas required around new public drinking water supply
wells also are common in many state public water supply statutes and
regulations.

Administrative Considerations

Traditional engineering standards require monitoring and enforcement to ensure compliance, and this can be problematic where sophisticated standards are imposed on a myriad of smaller point sources.
This is particularly the case where state and local resources may be
constrained or otherwise limited. For example, it is one thing to
require sophisticated engineering for a hazardous waste management
facility in a community, but quite another to require equally onerous
standards for several hundred (or even thousand) underground storage
tanks in the same community. Moreover, as engineering sophistication
increases, reliability tends to decrease (which accounts for the reti-
cence of many local boards of health to accept package treatment or
alternative small-scale treatment facilities in unsewered areas where
septic effluent poses groundwater threats). Finally, annualized O
& M costs may increase drastically beyond the additional capital costs
associated with various engineering strategies.

Land use control and locational restrictions, on the other hand,
are fairly easy to carry out, enforce, and monitor since the basic
administration mechanisms of land use controls are already in place
in many communities. They may perhaps be the best approaches for
addressing the incremental risks posed by many smaller sources. If
reliance on engineering is still preferred (for example, if such approaches are required under federal programs), then some innovative
administrative mechanisms must be created to carry out the management
program on the state and local level. Several approaches can be
considered, including (1) the designation of "lead agencies" to coordinate permitting and enforcement requirements that cross-cut various
sources; (2) expanding the regulatory authority of existing agencies
to enable them to address a wider variety of sources; or (3) shifting
the monitoring and compliance responsibilities onto the shoulders of
the landowner or operator of the source by requiring periodic reporting as condition of permit approval.

THE CONTROL OF NONPOINT SOURCES

Nonpoint sources--typically situations where large amounts of
toxic materials are introduced into groundwater over a relatively
large area--pose the greatest management problems. The most common
of these nonpoint threats is probably agricultural chemical application (pesticides and fertilizers) that infiltrates into unconfined
aquifers in relatively well-drained areas in rural areas. In urban
areas, nonpoint contamination sources can include the infiltration
of contaminated surface runoff and infiltration of roadway deicing
salts. Atmospheric deposition and infiltration of contaminated
precipitation can also pose pervasive problems in some locations; the
"acid rain" problem is beginning to be addressed in international
treaties, but may still pose short-term groundwater pollution issues.

Nonpoint sources often are addressed under federal legislation that requires proper notification of risks (TSCA) and which requires proper training in application (FIFRA) for agricultural pesticide and fertilizer applications in rural areas (and in some suburban communities concerned about chemical applications to lawns, parks, and/or golf courses). The Clean Water Act, especially areawide water quality management planning under section 208 of the Federal Water Pollution Control Act amendments, has also addressed run-off management; the Clean Air Act is likely to address atmospheric deposition issues. Yet, this panopoly of federal legislation may not be directed towards groundwater protection as a program objective, leaving the management of these sources up to state and local agencies.

Nonpoint sources are characterised by several common features that impede their effective management. One characteristic is their pervasiveness: very large quantities of toxic materials can be introdued into groundwater as a result of the great extent of application of some of these substances over a large area. A related characteristic is that the cumulative effects of nonpoint sources on groundwater quality often are not known since many rural areas rely on on-site drinking water supply wells that do not have to be tested under the Safe Drinking Water Act as do public supply wells. Few states or communities engage in systematic ambient groundwater quality monitoring to the extent that contamination from these sources can be readily identified. A third distiguishing complication is that relatively little is known about the toxicity of some of these substances, especially at low concentrations. Some contaminants, such as chlorides introduced by road de-icing salts or nitrates introduced by fertilizer applications, apparently pose few risks except in high concentrations (as established by current SDWA maximum contaminant levels); other substances, often found at very low concentrations, such as synthetic organic chemicals used as pesticides or carried in runoff, currently are not even addressed in the SDWA's primary drinking water standards (though the phase-in of listed contaminants under the recent SDWA amendments may broaden the range of potential threats that must be addressed).

Nonpoint sources also pose problems because traditional management practices have been directed towards the protection of surface water supplies and not groundwater resources. Runoff control is a good example: considerable efforts in state and local 208 plans were directed towards protecting surface water quality by minimizing runoff through the imposition of engineering measures to encourage infiltration and uptake by plants. Sedimentation basins, for example, characterize this approach. But surface waters may break down contaminants more readily than groundwater because of higher ambient temperatures, greater aeration, and exposure to sunlight. Conventional wisdom in preventing runoff encourages infiltration into aquifers, where the threats posed by contamination may be longer in duration and raise more significant health threats than the absense of regulatory intervention.

Best Management Practices for Nonpoint Sources

Current "best management practices" for rural nonpoint sources tend to rely on source reduction as a management strategy. Efforts have been directed towards minimizing the total amount of contaminants introduced into infiltrating precipitation and snowmelt, rather than relying on engineering measures to mitigate the contamination potential. These measures include restrictions or guidelines addressing pesticide and fertilizer application, to minimize the agricultural chemicals required to obtain a given crop yield; practices can include conservation tillage, manure application restrictions(limiting application on frozen ground), as well as integrated pest management approaches to minimize the amount of pesticides required.   Application rates and amounts of agricultural chemicals can also be guided to address local soil, crop, and hydrological features to prevent over-application.

Urban pollution issues are more complex, in some ways.   Detention and sedimentation basins still make sense, provided vegetation uptake of contaminants is addressed so that plantings in these basins are routinely cut and the cuttings disposed of properly.   Runoff contaminant source reduction--typically associated with the control of point sources involving hazardous or toxic materials use and storage--can possibly also help prevent infiltration risks, as well as periodic street sweeping and the establishment of effective stormwater management facilities.   Roadway de-icing salt applications can also be minimized by carefully managing the amount, timing, and mix of salts being applied.   The use of "tight" sewers in areas that are specially susceptible to groundwater contamination can minimize exfiltration and leakage threats where sewer installations are above unconfined aquifers.   Effective air pollution control measures--including the establishment of vehicle inspection and maintenance programs--may help reduce risks posed by atmospheric deposition.

Implementation Issues

Engineering approaches--sedimentation basins and stormwater drainage systems--only become feasible when development density is high enough to justify the capital costs of providing and servicing these types of facilities.   This limits these approaches to urban areas where adequate public facilities already are in place, or to suburban areas, where such facilities can be required as development exactions as a condition of development approval.   Management becomes more difficult in rural areas, where stormwater management may consist of a vegetated roadside ditch (which becomes useless if culverts and vegetation are not maintained to prevent blockage).   Effective engineering strategies therefore probably depend most on a community's willingness to bear the O & M costs associated with a facility, in addition to sharing in the financing of the capital improvement initially.

Rural nonpoint pollution control is also hampered by a laissez-faire attitude towards environmental management; many rural communities will not favor the imposition of stringent regulations that are likely to affect traditional agricultural practices.   Public education may be the most effective strategy in such communities, particularly where a farmer's own on-site water supply well would be at greatest risk of pollution from improper management practices.   Some communities, such as Jefferson County, Wisconsin, have adopted regulations that address manure spreading on frozen ground or which impose limits on the amounts of manure that can be applied on different terrains at different times of year.   As groundwater protection is recognized as an important issue in more rural communities, similar programs might arise in other parts of the nation.

Future Trends in Nonpoint Source Control

A number of communities, such as Rock County, Wisconsin, are beginning to assess groundwater contamination threats posed by both point and nonpoint sources.   This is also the case in Nebraska, where local environmental districts have been charged with preparing groundwater plans for their areas, and in southeastern Minnesota, where a environmental organization, The Minnesota Project, has developed model ordinances to protect groundwater in karst regions. These rudimentary efforts suggest that control of nonpoint sources is likely to emerge as a significant management issue in many rural areas over the next decade.

What is arising from these initial management programs is an enhanced willingness of rural communities to undertake risk assessments in order to focus their management efforts on priority contaminants threatening groundwater supplies.   For the most part, these supplies are exempt from monitoring and reporting requirements imposed on public water supply systems under the federal Safe Drinking Water Act (and its recent amendments requiring states to develop wellhead protection programs).   The early results of these risk assessments suggest that nonpoint pollution source control will emerge as the most significant issue that will have to be addressed.   Statewide water quality monitoring efforts, such as the network established by the Illinois Water Survey, will likely accelerate these efforts as contamination from nonpoint sources is identified as part of a monitoring program that is partly targeted towards assessing agricultural chemicals in ambient groundwater.   Data generated by the New York State Commission on Water Resources for Long Island has already resulted in restrictions on agricultural chemical use by many communities in that region.   Although regulatory approaches currently are shunned by many rural areas, they may emerge as the most effective method of addressing nonpoint pollution sources over the next few decades.

Effective nonpoint management through regulatory approaches raise some special problems in rural areas, though, since such areas often do not have adequate administrative resources to fully carry out effective management programs.   This is particularly the case where local officials and administrators hold only part-time positions with the relevant governmental units that may be given administrative responsibility for groundwater protection.   In such cases, innovative administrative entities--created at higher levels of government--may have to be created to assist local governments in the development of these programs.   Special management districts, encompassing several political jurisdictions, with full-time staff is one innovation that may prove useful in such circumstances. The establishment of a multi-jurisdictional "circuit rider" program to provide technical assistance to local units of government is another option. Finally, the traditional mechanisms of technical assistance from state agencies and from university cooperative extension services may have to be enhanced to provide adequate resources to undertake a comprehensive groundwater protection program. Programs developed under the U.S. EPA's sole source aquifer demonstration program, authorized under the recent amendments to the Safe Drinking Water Act, may serve as a good vehicle for fashioning these new administrative mechanisms.

CONCLUSION

The control of point and nonpoint pollution sources by state and local governments will be shaped by the development of federal legislation and regulatory initiatives over the next few years. Local program development, especially, tends to be reactive, while state efforts are often directed by federal funding opportunities under environmental protection legislation. But new programs are emerging that address groundwater contaminants and pollution sources that "fall through the cracks" of the federal programs and which have been identified as posing groundwater contamination threats in some communities and regions. The fragmentation of many federal source control programs leaves plenty of opportunities for other units of government to coordinate or integrate their own management programs in a more comprehensive manner, in furtherance of state groundwater protection strategies or local groundwater protection (or even 208) plans.

To date, the "best management practices" developed under these programs have been firmly rooted in "best engineering practices," particularly with point sources (although stormwater management facilities also address nonpoint sources as well). The administrative burden associated with ensuring engineering compliance may force a general trend towards broader intensity and locational restrictions, if administered by local levels of government with relatively few resources available to them. The regulation of nonpoint sources in rural areas is also likely to emerge as a management trend over the next decade, also focusing on non-engineering "best management practices" that address source reduction rather than pollution management. The long-term success of these efforts, however, will depend on the administrative innovations that can be fashioned in order to monitor, inspect, and enforce the next generation of local groundwater protection regulations that will be developed over the next few years.

The Delineation and Management of
Wellhead Protection Areas

Ron Hoffer*

Abstract

The delineation and protection of a management
zone around specific wells or wellfields will undoubt
edly become more common as the 1986 Amendments to the
Safe Drinking Water Act are implemented by the States.
Both primary elements -- the delineation of the zone
and the management of activities that may impair the
quality of the ground-water supply, can be accomplished
to varying degrees of comprehensiveness. Zone delineation is based primarily on the hydrogeologic setting
of the area, and conceptually represents a best-avail-
able characterization of all or part of the "ground-
watershed" supplying specific wells or aquifer segments.
Some of the primary factors affecting the extent of
this zone are the degree of aquifer confinement,
hydrologic properties of the aquifer, impact of recharge
and boundary conditions, and well construction and
operating characteristics. On the management side,
State and local governments will consider the stringency
of needed regulatory and nonregulatory controls depending on the actual or perceived threat to the resource,
as well as the legal, financial and administrative
tools available to them.

Introduction

Ground-water management has been associated
historically with systems to ensure that pumping not
exceed available recharge or that scenarios for "mining"
the resource are met. Ground-water quality had most
often been addressed only in coastal areas, where
recharge well barriers were placed to hydrologically
separate naturally occurring saline or brackish water
zones from water-supply wells.

While the management of quantities of ground water
will continue to play a very important role in resource
management plans, much more emphasis is being directed
to quality-management considerations. The Amendments
to the Safe Drinking Water Act passed in June 1986,
provide the legal framework for a new tool in this
effort -- that of wellhead protection. The Amendments
give State government the primary responsibility for
*U.S. EPA, Washington, D.C.

143

establishing such management zones around public water
supply wells in their jurisdiction. An overview of the
technical and administrative framework of wellhead
protection programs is provided in this paper.

Major Elements of Wellhead Protection Programs

     The Safe Drinking Water Act Amendments provide a
general definition of the wellhead protection programs
which the States must submit to the Environmental
Protection Agency by June 1989. Section 1428(a)
covers the six major elements, summarized briefly as:

   ° Identification of administrative roles in program
     development and implementation

   ° Delineation of the wellhead area associated
     with each well, based on "all reasonably avail-
     able hydrogeologic information"

   ° Identification of sources of contaminants within
     the areas

   ° Description of technical assistance, source
     control, training, and other measures the State
     intends to take to protect the wells from such
     contaminants

   ° Inclusion of contingency plans should wells
     become contaminated

   ° Consideration of potential contaminants within
     the expected wellhead areas of new wells

     A key point is that while the States are required
by statute to develop such programs, the EPA is not
given authority to take over for those States which
fail to meet this statutory mandate. The only "hammer"
that the EPA has against States that either choose not
to participate, or submit an "inadequate" program, is
the withholding of grant money. EPA is developing
technical guidance, however, for States to use in
applying for grant funds, as well as guidance for each
of the six specific program elements.

Definition of Wellhead Protection Areas

     As mentioned above, one major element of the
program is the delineation of specific wellhead areas.
The Amendments provide a general, legal definition of a
wellhead protection area (WHPA) in Section 1428(e). A
key passage notes that WHPA "means the surface and
subsurface area surrounding a water well or wellfield,
supplying a public water system, through which contami-

nants are reasonably likely to move toward and reach such water well or wellfield." It is clear by the legal definition, that the protection being offered to the wellhead goes beyond ensuring the physical integrity of the casing, cement, piping and other parts of the well itself. Few would deny, of course, the importance of such near-field measures since many contamination incidents are directly tied to improper cementing, casing, etc. The Amendments imply, however, a concern for a much more extensive look at portions of the surface and subsurface recharge area supplying water and potentially, contaminants from more distant problem areas.

In some situations the wellhead area might be delineated based on a simple fixed radius of one to several thousand feet, where in others, computer models are used to delineate a zone reflecting several decades of travel-time away from the well. As will be noted in this discussion, different policy and technical considerations will necessitate different operational interpretations of the Amendment's legal definition. The States have considerable flexibility under the Amendments for choosing their own approach, though EPA will need to define some base level of adequacy to meet overall statutory goals.

Current Experience in Wellhead Protection

There are few State-wide programs in the United States aimed specifically at wellhead or wellfield protection. The most frequently cited examples are those of Vermont (Vermont Department of Water Resources and Environmental Engineering, 1983), Massachusetts (Massachusetts Department of Environmental Quality Engineering, 1983), and Florida (Florida Department of Environmental Regulation, 1986). Within these States as well as in States without a specific wellhead program, individual counties or communities have developed similar strategies, or modified a wider State-specific concept (e.g., Dade County, Florida, Camp Dresser and McKee, 1982; Cape Cod, Massachusetts, Gallagher and Nickerson, 1986; Rock County, Wisconsin, Zaporozec, 1986).

In each case, a different approach to operationally defining the boundaries is used based largely upon the locality's particular goal for water supply protection. To cite one example, Florida's program has the underlying goal of establishing a zone where high-risk contaminating activities are banned. Their concept of a 5-year time-of-travel (TOT) boundary from given wells implies that if contaminants are released from such

activities beyond the wellhead area itself, the minimum
5-year separation would provide a type of "buffer
zone" to allow time for corrective action to take
place. Such limits on the extent of recharge area
included in the WHPA is based on the practical consideration that managing the full extent of the recharge
area, in many cases tens of miles long, would be impractical.

The question of how long it takes for corrective
action to be effective and hence, how much of the
recharge area need be included, has focused much of
the public debate on Florida's program. Proponents
for 10-year (or greater) TOT zones in Florida argued
that more time is needed for effective remediation.
Proponents for a 1 or 2 year zone had more confidence
in corrective action since they felt that most "high-
risk" activities are now monitored. This group further-
more believed that such strong land use controls as
banning of certain activities or industries should be
severely limited.

In terms of pollutant source controls in the 5-
year zone, Florida is targetting to prohibit landfills
and wastewater discharges containing hazardous chemicals
above background levels. The State also seeks to
manage certain domestic wastewater discharges, stormwater
discharges, underground tanks, pipelines, and the
handling, storage, or use of hazardous chemicals by
industry. The program in Dade County, Florida, while
concerned with similar pollution sources, in addition
utilizes a strong zoning and in-field inspection program
to assure that regulations are met, and other problems
managed.

To cite another example, implementation in New
England seems to be more protective since in several
instances the goal is to manage the entire recharge area
under current and projected future pumping scenarios.
Since the glacial aquifers used for most public supplies
rarely exceed a few square kilometers in area, rather
than the many hundred of square kilometers typical in
Florida, a more protective goal in those cases is also
a more implementable one. Management strategies,
including permits, regulations, inspections, and zoning
changes, are often targetted to restrict housing/septic
tanks, hazardous waste disposal, landfills, and
pipelines.

Contrasting the limited State experience in wellhead
protection in the U.S. is the widespread use of such
programs in Europe. At least eleven European countries
have some form of national protection zone concept
(van Waegeningh, 1985). The two countries with the

most extensive experience in this area are West Germany
and the Netherlands. Common to both "water protection
area" schemes are an immediate well zone to protect
wellfield hardware itself from spills and leaks, with
a secondary zone of 50 or 60 days TOT to ensure reten-
tion of microorganisms in the subsurface. Each country
then applies "water protection" area zones which are
most comparable to the WHPA boundaries we may see in
the U.S. In West Germany this zone extends to 2 kilometers from the well bore (along flowpaths, and until
aquifer boundaries are reached) whereas in the Netherlands the protection zone extends to 10 and 25 year
travel-times. Under average hydrogeological conditions
in the Netherlands, these TOT boundaries correspond
respectively to approximately 800 meters and 1200
meters from the wells. Within these protection areas
there are severe restrictions against such activities
as waste disposal sites, the transport and storage of
hazardous chemicals, high-risk industries, rock and
aggregate quarrying, wastewater disposal, feedlots and
pesticide storage, and application of moderate to
highly leachable pesticides. Finally in both countries,
an outermost zone is drawn to the recharge area boundaries, though pollutant source controls are somewhat
less stringent.

Hydrogeologic Considerations

        It can certainly be argued that the primary issues
to be faced by State and local governments in their
wellhead protection programs will be administrative and
policy in nature. If the program is to be a success,
however, it is in part dependant on the technical
underpinnings of the chosen approach. The range in
hydrogeologic settings and the variabilities in technical
data at specific well sites has been recognized in the
Amendments as key reasons for EPA providing a broad
spectrum of approaches in guidance to the States. Many
of the major technical questions on hydrogeologic aspects
of WHPA delineation fall into two broad categories --
criteria for boundary line determinations, and methods
to calculate the in-field expression of these criteria.

        There are at least five major criteria on which to
base the protection area boundary -- distance from the
well, the amount of drawdown in the water table or
potentiometric surface that the pumping can cause, the
time of travel to the well, the physical boundaries of
the aquifer and/or local recharge area, and the amount
of physical area needed to ameliorate contamination so
as to maintain water quality below drinking water
standards or action limits. Accompanying these criteria
are specific thresholds which define the actual boundary.

Example of such thresholds include:

° Distance -- 300 meters (proposed; Nebraska)
           -- 2 kilometers (West Germany)

° Drawdown -- 0.08 meters (some Florida counties)
           -- 0.03 meters (Cape Cod, MA)

° Time of Travel -- 5-years (Florida)
                 -- 10 and 25 years (Netherlands)

° Assimilative Capacity -- Meet "Action Level" for
                           nitrates (suburbanizing
                           areas of Northeastern
                           U.S.)

° Physical Boundaries -- till/stratified drift
                         contact (Vermont)

     Many localities with wellhead programs incorporate
several different criteria. The Netherlands, for
example, utilizes distance for its innermost zone, time-
of-travel for its key implementation zones (50 days;
10 years and 25 years), and physical boundaries for its
outermost zone. There are also varying interpretations
of certain criteria. "Physical boundaries" in the
case of smaller alluvial aquifers, will mean an unchange-
able contact between highly permeable aquifers and
much less permeable surrounding formations. The interface between these two materials is a very effective
boundary to flow. In other cases, the boundary may be
somewhat more hydraulic in nature, and be subject to
change if there are major shifts in pumping patterns.
A stream which serves as a discharge boundary to aquifer
egments on either side of it, for example, may no
longer be effective in this function should extensive
pumping from new wellfields on one side induce a large
quantity of underflow from across the river.

     There are other questions which will be raised
when choosing candidate WHPA criteria including:

° "Protectiveness" of Criteria -- The greater the
  threshold (e.g., 50 years travel-of-time versus
  5 years travel-of-time; 2 kilometers versus 0.2
  kilometers), the greater the protection, except
  if limited financial and managerial resources
  imply less overall program effectiveness in the
  larger area.

° Relevance to Contaminants of Concern -- Almost
  all current programs base time-of-travel criteria
  on uncontaminated ground-water characteristics,
  conditions which lead to considerable variations
  in actual contaminant arrival times. This is

more of a problem in predicting effects in
shorter rather than more lengthy wellhead areas.

° Appropriateness to Semi-Confined and Confined
  Aquifers -- Although WHPAs are most often applied
  to water-table aquifers, the Amendments do not
  "exempt" specific aquifer settings from the
  program. Some States may choose to use different
  criteria for extensively confined aquifers to
  prevent, for example, undesireable penetration
  of that confinement. States will also need to
  know when a confined aquifer is "leaky enough"
  to require efforts similar to water-table
  settings.

° Appropriateness to Other Aquifer Conditions -- A
  10-year time-of-travel criteria is irrelevant
  for channelized karst areas, where physical
  delineation of conduit channels (e.g., by dye
  tracing) is far more relevant. Fixed distance
  criteria have more appeal in large, semi-confined
  sandstone aquifers with many small wells, due
  to the lack of site-specific pumping data. The
  same approach in a small, highly productive and
  highly utilized alluvial aquifer would be less
  appropriate, except for a first-step approxi-
  mation.

Delineation Methods

     Except for the "distance" criterion, one which
normally requires a ruler or map scale to implement,
some technical method is required to translate the
remaining criteria to a line/area on a map. While
there is considerable use of multiple methods in given
State or local programs, the methods generally comprise
seven categories:

° Fixed Radius -- Where a circle is drawn at a set
  distance around the well. Usually there is
  little technical basis to the distance chosen
  though it does offer a measure of protection
  over and above that without any WHPA.

° Calculated Fixed Radius -- Here a simplified
  radius (i.e., no variation in shape as a function
  of water-table slope or variations in aquifer
  properties) is used, but the basis is tied to
  some in-field properties such as averaged aquifer
  thickness, porosity, and well pumping
  rate. This approach was chosen by Florida to
  implement its 5-year time-of-travel criterion.

° Simplified Variable Shapes -- This is based on
  the development of a fixed set of different

shaped protection areas based on summary analyses
of major hydrogeologic conditions and pumping

patterns in a given area. The Southern Water
Authority (1985) in England, for example,
developed a set of several shapes ranging from
small circles (low pumping, little slope in
watertable) to broader ellipsoids (higher
pumping; characteristic water-table slope and
direction).

° Analytical Flow Model -- Probably the most widely
used approach in the Northeastern States, it
determines the area of contribution of a well
through application of an analytical flow equation.

° Numerical Flow and Transport Model -- Typically
employed where greater accuracy is desired, and
data and cost requirements are met, these
computer models offer potentially the most
accurate delineation method. The county programs
in southern Florida commonly use 2-dimensional
models (Camp, Dresser, and McKee, 1982), while
a test of a 3-dimensional model is planned for
Cape Cod.

° Geologic and Geomorphic Mapping -- In near-
surface aquifer settings, where physical boundaries are a primary delineation factor, geologic
and geomorphic mapping techniques will be an
effective approach from both cost and economic
perspectives. These boundaries will match
topographic/drainage divides in some settings.

° Miscellaneous Methods -- A growing list of more
specialized techniques includes natural and
induced tracing procedures, and geophysical
surveys of buried aquifer boundaries.

Each of the methods is accompanied by a set of
characteristic features, advantages, and disadvantages.
The most sophisticated method, numerical modeling, has
the potential for accomodating localized recharge,
discharge, variable boundaries, aquifer properties, and
many other features. The relative high cost of full
model development and verification may be justified if
strict control strategies (with potential high economic
impacts) are proposed. Other States may propose an
intermediate test (e.g., calculated fixed radius or
analytical model) since it can offer a good degree of
protection at a moderate degree of investment, and in a
rapidly implementable mode. Still others may find the
simplicity and administrative ease of a fixed radius
approach, and counteract the possibility of "underprotec-

tion" by choosing a more extensive threshold.

Conclusion

In its guidance and support documents on wellhead
protection area delineation and management to the
States, the Environmental Protection Agency will
evaluate these and other approaches to protection so
that States may begin their program from a common base
of experience. As the wellhead protection program is
implemented, new directions in both the delineation
and management aspects of the wellhead program will
undoubtedly emerge. The Agency will continue to play
a major role in evaluating and communicating this
information so as to make this tool for ground-water
protection an even more effective one.

Disclaimer

While the thoughts presented in this paper were
gathered from the author's experiences in the Office of
Ground-Water Protection at EPA, they should not be
construed to represent official Agency policy or
direction on implementation of the Safe Drinking Water
Act Amendments. Such policies will be officially
released by EPA along with its guidances and grant-
related application packages.

References Cited

Camp, Dresser, and McKee, Inc., 1982. Wellfield Travel
Time Model for Selected Wellfields in Dade, Broward,
and Palm Beach Counties, Florida. Fort Lauderdale, FL.

Florida Department of Environmental Regulation, 1986.
Proposed Revisions of Chapter 17-3, Florida Administrative Code, Class G-1 Ground Water, Tallahasee, FL.

Gallagher, T. and S. Nickerson, 1986. The Cape Cod
Aquifer Management Project: A Multi-Agency Approach to
Ground-Water Protection. Proceedings of the National
Water Well Association, Third Annual Eastern Regional
Ground-Water Conference.

Massachusetts Department of Environmental Quality
Engineering, 1983. Groundwater Protection Strategy.

Southern Water Authority, 1985. Aquifer Protection
Policy. West Sussex, United Kingdom.

van Waegeningh, H.G., 1985. Protection of Ground-Water
Quality. pp. 111-121 and 159-166. In Matthes, G., et.
al., editors, Theoretical Background, Hydrogeology and
Practice of Ground-Water Protection Zones. Inter-
national Association of Hydrogeologists, volume 6,
UNESCO, Hannover, West Germany.

Vermont Department of Water Resources and Environmental Engineering, 1983. Vermont Aquifer Protection Area Reference Document, Montpelier, Vermont.

Zaporozec, A. (editor), 1985. Ground-Water Protection Principles and Alternatives for Rock County, Wisconsin. Wisconsin Geological and Natural History Survey, Madison, Wisconsin.

EMERGENCY RESPONSE

Alexander J. Fazzini and Art Rosenbaum
(ASCE Member)

Most hazardous-material spills are caused by transportation accidents, incidents at a fixed facility, or by mother nature. When an incident occurs, the emergency-response contractor develops an approach in compliance with applicable federal, state, and local regulations regarding such matters and in cooperation with the responsible party. The response contractor must be qualified and be capable of providing immediate service.

A spill normally has five phases; namely, the initial emergency response, the site assessment, the development of a remedial action plan, the implementation of the remedial action plan, and the post cleanup monitoring. The primary focus of the emergency-response phase is to mitigate any immediate threat to the health and safety of the surrounding community and to the personnel performing the work. The approach during the emergency phase should limit the extent of contamination to the surrounding environment. The more recovery of product during the emergency phase to minimize environmental damage, the more likely the restoration effort will be simpler and less costly.

The paper will explain in further detail the concepts described above, including the type of planning and resources necessary to respond and remediate hazardous-material spills. It will explain the implications and interrelationships of each of the project phases with regard to the sources, pools, and plume of contamination. Levels of protection, work zones, methods of containment, product recovery, ground-water recovery, on-site treatment, and air monitoring will be examined. A case study will be presented to demonstrate the identified concepts.

Alexander J. Fazzini   Art Rosenbaum
Regional Manager   Technical Manager
O.H. Materials Corp.   O.H. Materials Corp.
Windsor Industrial Park   Windsor Industrial Park
P.O. Box 41   P.O. Box 41
Windsor, NJ 08561-0041   Windsor, NJ 08561-0041

Accidents happen. When an accident occurs involving hazardous materials or waste, standard emergency procedures do not apply. Spills, whether caused by transportation accidents, incidents at a fixed facility, or by mother nature, require a response from a government agency or a contractor who is well-versed, not only in emergency procedures, but in health, safety, and cleanup practices as well. Specialized equipment may also be required to address the spill. An environmental-services contractor must have trained personnel, specialized equipment, and most importantly, experience to approach and manage spills in a manner which will address all applicable federal, state, and local regulations.

When a hazardous-materials spill occurs, immediate response is warranted. This response, by a qualified hazardous waste-cleanup contractor, takes the form of planned remedial actions (although the time frame is greatly compressed, the emergency response still will be executed in a planned and practiced fashion). Many factors, such as applicable federal, state, and local regulations and immediate threat to the public and environment are to be taken into consideration by the contractor. The contractor must be well-versed in handling these factors as well as in the implementation of appropriate health-and-safety measures for personnel performing the cleanup. A spill typically has five phases; initial emergency response, site assessment, development of a remedial action plan(s), implementation of remedial action plan(s), and post cleanup monitoring.

To formulate an initial response, information is needed concerning the nature of an emergency, including the type, quantity and physical nature of the material spilled, the incident itself, and location of the spill. A contractor will receive this information and provide an appropriate response concerning proper protective equipment required by personnel, equipment and manpower resources required on site, and notification of all appropriate federal, state, or local agencies. Levels of personal protective equipment are divided into four categories: A, B, C, and D (Table 1). A is the highest level, while D is the lowest. In many cases, information initially received concerning protective level requirements must be confirmed or corrected by the contractor's response manager.

TABLE 1

LEVELS OF PROTECTION

Level A: Should be worn when the highest level of respira-
            tory, skin, and eye protection is required.

Level B: Should be selected when the highest level of res-
            piratory protection is needed, but a lesser level
            of skin protection.

Level C: Should be selected when a lesser level of respira-
            tory protection is needed, skin protection is the
            same as Level B.

Level D: Basic work uniform for support zone (office,
            decontamination) where no contamination exists.

The primary focus of the initial emergency response
phase is to mitigate any immediate threat to the health and
safety of the surrounding community and environment. Upon
arrival at a spill location, a response manager will evaluate all necessary factors and develop a plan of action.
This plan of action will address such tasks as product containment and recovery and setting up work zones. The response manager will also assess the spill to determine the
equipment necessary to contain and recover as much product
as possible. This effort will minimize environmental damage
and reduce long-term restoration efforts.

A generalized scheme of work zones is illustrated in
Figure 1. The exclusion area is the innermost area and is
considered contaminated or hot. Within the exclusion area,
prescribed levels of protection must be worn by all entering
personnel. The exclusion area boundary is established initially based on the type of contamination, initial instrument readings, and safe distance from any potential exposure.

Between the exclusion area and the support area is the
contamination-reduction zone. The purpose of this zone is
to provide an area to prevent or reduce the transfer of
contaminants which may have been picked up by personnel or
equipment returning from the exclusion area. All decontamination processes occur in this area.

The boundary between the contamination-reduction zone
and the exclusion area is the hot line. An access control
station must be established at this point to control flow
of personnel and equipment between contiguous zones and to
ascertain that all procedures established to enter and exit
the zones are followed. Entrance into the exclusion area
requires wearing prescribed personnel protective equipment,

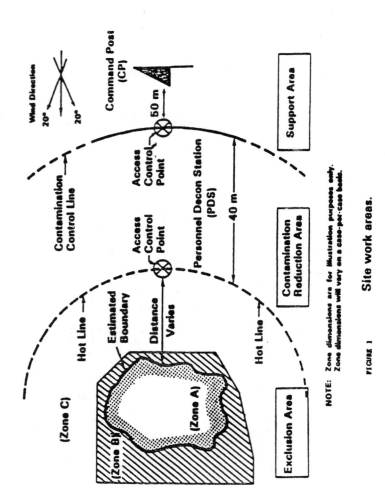

NOTE: Zone dimensions are for illustration purposes only.
Zone dimensions will vary on a case-per-case basis.

FIGURE 1   Site work areas.

which may be different than equipment requirements for working in the contamination-reduction zones. At a point close
to the hot line, a personnel and/or equipment decontamination
station is established for those exiting the exclusion area.
In some, cases, another decontamination station is required
closer to the contamination control line for those working
only in the contamination-reduction area.

Use of a three-zone system of area designation, access
control points, and exacting decontamination procedures
provides reasonable assurance against spreading the contamination. This control system is based on a worst case situation. Less stringent site control and decontamination procedures may be utilized based upon more accurate information
on the types of contaminants involved and the hazards they
present. This information is obtained through air monitoring, instrument survey, and technical data concerning the
characteristics and behavior of the material present. Site
control requirements can be modified based on the specific
situation.

Once the initial phase of the cleanup effort is complete, site assessment is performed to determine further
remedial actions which may be necessary. To make this
assessment, investigations into the extent of soil and
ground water contamination are needed. A drill rig (either
hand held or truck mounted) is used to determine the extent
of the pool or plume of contamination. Initial soil sampling
can indicate not only the extent, but also the direction the
contamination is moving. This preliminary information will
help in determining where soil must be excavated and/or
monitoring and recovery wells can be placed to ensure maximum effect. This remedial action plan will take into account
the cleanup levels and effort necessary as required by the
appropriate federal, state, or local codes. This remedial
action can take the form of soils cleanup, ground water
recovery and treatment, or long-term monitoring with no
cleanup effort.

An example of the concept discussed above is a case
where a tanker truck carrying gasoline rolled over and
spilled its contents into an area where municipal drinking
water wells were located. The initial response was to contain and collect as much gasoline on the ground and in a
nearby stream as possible. The imminent danger of the ex-
plosive nature of the gasoline was thus removed. The soil
in the area was sandy and the ground water table was determined to be a few feet below ground surface. The next step
was to determine the extent of contamination, horizontally
and vertically. The ground water was also analyzed for
the dissolved products of gasoline.

Once the extent of contamination was known, a treatment system for soil and ground water decontamination was

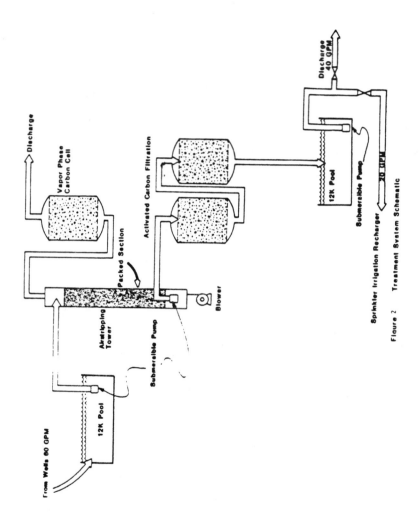

Figure 2   Treatment System Schematic

designed and implemented. The design for this system also
took into account the cleanup criteria required by the appropriate regulatory agency. Figure 2 shows the treatment
system for this example. To alleviate the need for an air
discharge permit, a vapor phase carbon cell was added to
the air stripping tower. Treated water was then flushed
onto the spill area to wash the contamination which adsorbed
onto the soil. Initially, contamination for the dissolved
product of gasoline was in the 30,000 parts per billion
(ppb) range. Within a few months, analyses indicated contamination reduced to the 100 ppb range.

When to Remediate:
Remediation or Reparations

Dr. K. C. Bishop III[*]

The 1986 reauthorization of the Comprehensive Environmental Response, Compensation, and Liability Act (Superfund) forcefully stated Congress' desire to see our nations hazardous waste sites cleaned up. There is no question the public, state government, environmental organizations and industries all support this objective. The debate is no longer over funding levels, judicial remedies or taxes. As we get on with the program the focus of the debate will be, What is a remedial action? When and why are remedial action required? and, What are the public policy goals for such actions?

Introduction

The Environmental Protection Agency has identified 25,000 sites as potentially covered by Superfund. The National Priority List (NPL) now stands at 888. It is estimated the number of reported sites may grow to over 40,000 and the NPL to 12,000. Against this background the public perceives that new sites are continuously being generated at an alarming rate.

The actual situation is far less appalling than the above facts suggest. First, the huge number of potential sites are the result of the overly broad definition of reporting requirements under CERCLA Section 103(c).

"Within one hundred and eighty days after enactment of this Act, any person who owns, operates or who at the time of disposal owned or operated, or who accepted hazardous substance for transport and selected, a facility at which hazardous substances are or have been stored, treated or disposed of shall . . . notify the Administrator of the EPA."

Taken literally, this law required anyone who had ever spilled any hazardous substance at any location to report that location to the EPA. Nevertheless, EPA has conducted a preliminary assessment on well over 19,000 of these sites. Of these, it appears only 1,450 are likely to represent a threat to man or the environment significant enough to be considered for the NPL. (McCleoud, 1986.) Moreover, under existing laws and with our current knowledge new sites are not being generated at a significant rate.

[*]Policy Coordinator, Environment and Health Issues, Chevron U.S.A., 575 Market Street, San Francisco, California 94105.

Current legal requirements and technology will prevent the kinds of mistakes which led groundwater contamination in the past. Under RCRA all hazardous waste storage, treatment and disposal sites are carefully scrutinized. Indeed, land disposal is subject to ever more stringent requirements and prohibitions just to protect groundwater. Underground tanks will soon be subject to detailed requirements -- not only for new tanks but also to monitor and upgrade existing tanks. Pesticides which have reached groundwater are now the subject of "special reviews" by the EPA Office of Pesticide Programs. All of these programs and more have come from our understanding of how man's activities cause groundwater contamination. New degrees of care are preventing the majority of the problems before they occur -- and prevention is the real goal in protecting groundwater resources.

Only six sites on the NPL have been identified as having clean ups completed. (Note: an additional eight sites are proposed for de-listing.) This has resulted partially from the very real technical, legal and political constraints to remedial action. The reauthorization of CERCLA implicitly recognizes these constraints. The language balances the cries of frustration that "clear cut standards are needed" against the need for "case-by-case, cost-effective solutions." Additionally, many of the sites have been 99% remediated but the agency maintains a listing on the NPL to assure oversight until the work is complete. In fact, the 2-year debate over reauthorization ultimately acknowledged the existing process which is, and will continue, to work to clean up these sites.

While the EPA has ostensibly moved slowly on cleaning sites on the NPL, the states and private industry have been moving quickly to address those other sites of potential concern. For example, the state of California projects 201 sites which will undergo clean up in fiscal year 1986-87 (Calif. DHS, May 1986). Ninety-eight of these sites are on the NPL. New York reported 21 that were removed from their list last year (Sterman, 1986). Other states are acting to address their immediate concerns. Simultaneously, private corporations and companies have taken on the responsibility to clean up their former contaminated sites -- even though their actions were entirely legal at the time. Ms. Michele Corash, former General Counsel for EPA, testified before Congress (Superfund Reauthorization, 1985).

"It's revolutionary in the sense that we are talking in, most instances, about parties which behaved completely lawfully, which in many instances, behaved in accordance with, or at the direction of government agencies in disposing of wastes which have now turned out to create a problem. Throughout the country, parties, including those who are engaged in litigation, have stepped forward and accepted the responsibility for cleaning up these wastes."

How have the decisions for these cleanups been made? An initial reaction might be that they have to fulfill the legal responsibility to clean up sites of contamination. However, experience suggests more than this simple answer, it is a complex set of decisions based on the unique circumstances at each individual site. The remainder of this paper discusses the key issues -- legal requirements, public/political concerns and potential liability. Each of these elements plays an important role as each cleanup develops (See for example Calif. DHS, May 1986).

Discovery and Assessment

The initial discovery and/or assessment of the site has an important impact on which inputs will prove important in making a remediation decision. If the site is discovered, assessed and reported to the requisite agency by the responsible party then technical considerations will weigh more heavily -- rightfully or wrongfully -- on the conduct of remediation. In contrast, if a site is brought to the attention of the agency by homeowners experiencing adverse health effects, the decisions will be based on other considerations. Between these extremes lie a continuum of possible degrees of many differing inputs.

The discovery/assessment phase should provide a technical assessment of:

1. what is at the site;
2. what substances are of concern; and
3. possible paths of migration from the site.

At most locations, a site history gives an overview of the types of substances likely to be present. From this list the substances of the most concern can be selected. This process is an iterative process which also depends upon the likely routes of exposure. For example, different sets of chemical properties are important if you are concerned about air or groundwater exposure pathways. In most cases, groundwater is the most likely pathway for migration. Fortunately, the list of substances likely to be important belongs to a limited number of chemicals. Table I lists many of the substances which have been found frequently in groundwater. These substances, because of their chemical and physical properties, are the most likely focus of a remedial action involving groundwater.

TABLE I

FREQUENT CONTAMINANTS OF GROUNDWATER*

Tetrachloroethylene
Trichloroethylene
Dibromochloropropane
Chloroform
1, 1 - Dichloroethylene
Carbon Tetrachloride
Benzene
Atrazine
Aldricarb
1,1,1-Trichloroethane

*Compiled from California DHS, April, 1986, K.C. Bishop, 1986 and Universities Associated for Research and Pathology, 1985.

Also during this initial phase, significant point sources of contamination are removed or closed. For example, drums are sent to an incinerator or a secure land fill; obvious "hot spots" of soil contamination removed; ponds drained, tanks closed, etc.

At this point in the remediation of a contaminated site: the site has been identified to the responsible agency; the obvious sources of contamination have been removed; the substances of concern have been identified; and the probable pathway(s) of migration and/or exposure identified. The next step is to begin an evaluation of groundwater remediation.

Remedial Actions

Once the source has been controlled, the decisions on remediation become increasingly complex. Often the participants in the decision will have competing goals. The approach outlined here is substantially that adopted by the Chemical Manufacturers Association in December of 1987.

The primary objective of remedial actions should be to preserve the beneficial use of the resource. Remediation should not be used as a method to extract reparations from the site operators -- such resources are better spent on prevention of future groundwater contamination.

Standards should be used to ascertain that the use will not be impaired. The possible impairment of use should be judged at the point of use. Costs should be considered in choosing among the possible remedial actions that allow continued beneficial use of the groundwater.

There are undeniably a tremendous range of potential remedial actions available. Some of these can be extremely expensive. Table II provides a recent assessment of the costs of some of the possible groundwater remedial actions. Obviously, with such extraordinary costs the focus must be on assuring the use is not impaired . . . not just cleaning up groundwater for cleanups sake.

TABLE II

COSTS OF GROUNDWATER REMEDIAL ACTIONS[1]

Action	Cost[2]
Replace Water Source	$1MM/gpd
Cap and Monitor	$150M + $30-50M/year
Water Removal and Treatment	$2-3/1,000 gallons
Aquifer Rehabilitation	$20MM+
Wellhead Treatment	$150/year per household
Encapsulation	$5-15/square foot

[1] Adapted from CMA draft, September 22, 1986.
[2] M = thousand; MM = million.

The contamination of water without beneficial use should not be allowed. Rather, we should use restraint when correcting past mistakes. Remedial action should not be automatic, it is not a method of extracting reparations. This punishes the responsible party (who may have performed totally legally) but does nothing to protect groundwater. We are better served by using available private and public funds and manpower to protect the beneficial uses of the groundwater resource.

The most obvious beneficial use is drinking water. Groundwater is a source of drinking water to 50% of the population. About 95% of all rural residents depend on groundwater for their daily water supplies. No single issue is more fraught with emotion and misperception than drinking water/groundwater contamination. As such, it is the best example of how decisions are made on remediation.

A reading of the popular literature would suggest this is already an important public health problem. However, most hazardous waste sites have not been clearly associated with human health concerns. Indeed, in 1985 the UAREP report summarized the knowledge about health effects from waste sites: "Experience to date is limited, but the data do suggest that human exposures, in general, have not been of sufficient duration and concentration to have resulted in observable long-term health effects. Based on our current knowledge, chronic health effects from long-term, low-level exposures would be expected to affect only a small percentage of those exposed."

Remediation of chemical contamination of drinking water begins with the assessment of possible use. This assessment should consider the questions:

o Is the aquifer now used for drinking water?

o Is the natural or background purity suitable for drinking water?

o Are there economically recoverable quantities of groundwater given the other sources available?

Water which is used for drinking water will necessitate some form of remedial action. If the facts suggest that development of the groundwater as drinking water supplies is likely, sometime in the future, then some form of remedial action will be required. Conversely, if it is improbable that economic quantities of sufficient quality are or will be available from the aquifer -- then there should be less emphasis on remediation. In such a case, no current or beneficial use is impaired. In short, all groundwater is not drinking water. Obviously, such an assessment requires a careful case-by-case analysis of the specific groundwater and its aquifer which is contaminated. Drinking water is the primary concern, but other uses, e.g. irrigation, cooling water, are equally important.

A decision that requires no remediation often flies in the face of de facto public policy, public concern and liability questions. Therefore, too frequently, remedial actions are demanded: "In many cases in the Silicon Valley; for example, stringent cleanup levels have been used to clean up hydrographic units not used for drinking water and where substantial alternative drinking water supplies exist." (DeVries, April 1986)

What has driven these decisions away from the technical answer? For example, the groundwater contamination in the Santa Clara Valley of California has been of great concern. Professor Bruce Ames has stated, "Thirty-two of the thirty-five wells with chlorinated solvent contamination are safer than tap water." Nevertheless, huge sums have been spent to clean up the groundwater because the public perceives a potential problem. Under the regulations of the National Contingency Plan, the public must have an opportunity to comment. The agencies are driven by this expressed concern.

The reauthorized superfund legislation will lead to even further input from the public. First, public information and comment are required before any remedial action. Second, grants will be available to "any group of individuals which may be affected by a release or threatened release at any facility on the National Priority List." Finally, any person with an interest may intervene when they claim that the remedial action will impair their interest.

Proposition 65 in California will allow any citizen to bring a lawsuit to prevent contamination of any underground source of drinking water. In total, the public will have an even bigger say in future cleanup actions -- potentially raising the spectre that only total removal or destruction will be satisfactory. The "not in my backyard" sentiment, evident in siting new RCRA waste disposal facilities, suggests the public will always demand remediation.

Additionally, today no company wants its products or waste in drinking water -- or potential drinking water. The recent settlement of the "Woburn" case points to the high costs of litigation of such issues. Remedial action will be cheaper than potential liability if the substances are in drinking water. Moreover, no company wants a reputation as a polluter. Even unused zones have been the subject of remedial action when reasonable cost effective means were available for clean up. Modern corporations desire to be perceived as responsible members of society. This desire will also often lead to the decision to remediate beyond the technical considerations.

If remedial action is needed, the search should be for the most cost-effective solution to preserve the beneficial use. Such alternatives run from replacing the water source to aquifer rehabilitation. For example, if rural wells are completed in a shallow zone, the entire zone could contain low levels of a single pesticide. It is impractical to attempt to clean the entire water bearing zone -- if it's possible to either supply a filter to the existing well or provide another deeper well (Holden, 1986). Conversely, if there is a single extremely concentrated point source of contamination, complete removal would prevent widespread contamination. The actual selection depends on a case by case analysis of the possible impairment of use.

Standards and criteria will continue to be used to determine the impact of chemical contamination on beneficial use. However, the appropriate standard should be used in the context they were developed. Most standards and criteria generally are based on long-term low-level exposure. The objective of the remediation is to interrupt this long-term exposure and restore the water to a quality suitable for use. Therefore, so-called ambient groundwater standards should not be used independent of the site specific analysis discussed above.

Conclusion

The focus of groundwater in the coming decade must be on protection. Decisions to remediate and the method of remediation should focus on preserving the beneficial use of the aquifer. Cleanup of groundwater must be viewed in this way, for cleanup is not a punishment -- but a responsibility to future users of the water.

REFERENCES

Ames, B. N., Testimony to California Senate Committee on Toxics and Public Safety Management, Sacramento, CA, November 11, 1985.

Bishop, K.C., Proceedings of the U.S. Conference on Irrigation and Drainage, Phoenix, Arizona, October 23 and 24, 1986 (in press).

California Department of Health Services, California Site Mitigation Decision Tree (Draft Working Document), June 1985.

California Department of Health Services, Expenditure Plan for the Hazardous Substance Cleanup Bond Act of 1984, (Revised) May 1986.

California Department of Health Services, Organic Chemical Contamination of Large Water Systems in California, April 1986.

Chemical Manufacturers Association, Program for State Groundwater Management, Chemical Manufacturers Association, Washington, D.C., January 1987.

DeVries, Johannes, (ed.) Proceedings of the Fifteenth Biennial Conference on Groundwater, Water Resources Center, University of California, Davis, CA, April 1986.

Environmental and Energy Study Institute: Special Report, Groundwater Protection: Emerging Issues and Policy Challenges, EESI, Washington, D.C., October 23, and 27, 1986.

Federal Register, "National Oil and Hazardous Substances Pollution Contingency Plan", Volume 50, No. 224, November 20, 1985, pages 47912-47979.

Federal Register, "National Oil and Hazardous Substances Contingency Plan: National Priorities List Update", Volume 50, No. 251, December 31, 1985, pages 53448-53452; Volume 51, No. 45, pages 7934-7935.

Environmental Protection Agency, Pesticides in Groundwater: Background Document, Office of Groundwater Protection, May 1986.

Gordon, Wendy, A Citizen's Handbook on Groundwater Protection, National Resources Defense Council, New York, NY, 1984.

Holden, Patrick W., Pesticides and Groundwater Quality: Issues and Problems in Four States, National Academy Press, Washington, D.C., 1986.

Marshall, Eliot, "Woburn Case May Spark Explosion of Lawsuits", Science, Volume 231, pages 418-420, October 24, 1986.

McLeod, Don, "Waste Not Pay Anyway: Most Firms Foot Cleanup Bill", Insight, November 17, 1986, pages 18-19.

Pye, Veronica, I., Patrick, Ruth and Quarles, John, Groundwater Contamination in the United States, University of Pennsylvania Press, Philadelphia, PA, 1983.

Sterman, David, et al., Joint Legislative Commission on Toxic and Hazardous Wastes, In Search of Progress: A Critical Review of the New York State Hazardous Waste Cleanup Program, 1986.

Superfund Amendments and Reauthorization Act of 1986, Conference Report to the House of Representatives, Report 99-962, October 3, 1986.

Superfund Expansion and Protection Act of 1984, Hearing, July 25, 1984, before the House of Representatives Ways and Means Committee, Serial 98-95.

Superfund Reauthorization: Judicial and Legal Issues, Oversight Hearings, before the House of Representatives Committee on the Judiciary, Serial 19, July 17 and 18, 1985.

Universities Associated for Research and Education in Pathology, Inc., Health Aspects of the Disposal of Waste Chemicals, Bethesda, MD, February 1985.

Biological Treatment of Trichloroethylene In Situ

John T. Wilson, Sam Fogel and Paul V. Roberts*

Abstract

Trichloroethylene and related compounds such as cis- and trans-1,2-dichloroethylene and vinyl chloride are common contaminants of ground water in industrial areas. In oxygenated ground water, these compounds are generally resistant to biodegradation. It has been recently shown that these compounds can be cometabolized by bacteria that oxidize gaseous hydrocarbons such as methane or propane. The hydrocarbon-oxidizing bacteria excrete the corresponding epoxide, which rearranges or hydrolyzes to other compounds. Then these products are degraded to carbon dioxide by other naturally occurring bacteria. This cometabolism forms the basis for an innovative biotechnology to reclaim ground water. Currently, a field study of trichloroethylene degradation is being carried out in a shallow semiconfined aquifer in the Santa Clara Valley of California. This field study has good control on dilution of contaminants due to dispersion, and on removal of trichloroethylene through nonbiological processes, which allows an accurate determination of trichloroethylene removal through biodegradation. Field studies are also planned for biodegradation of dichloroethylene or vinyl chloride in a contaminated aquifer somewhere on the East Coast.

Introduction

Contamination of aquifers with trichloroethylene and related compounds is widespread. These compounds are used as degreasers in the electronics industry and in the manufacture and maintenance of fine machine parts such as aircraft engines. Our national commitment to protect ground water quality has resulted in numerous attempts to reclaim aquifers contaminated with these compounds. The most common approach involves extracting the ground water by pumping and subsequently treating the water at the surface by carbon adsorption or air stripping. Carbon adsorption is expensive and air stripping without adsorption merely transfers the contaminant to another environmental medium. Because in situ treatment circumvents these difficulties, it is an attractive alternative to conventional practice. This report summarizes our progress on one particular approach--the biological degradation of the contaminants to carbon dioxide and chloride ions by bacteria that occur naturally in aquifers. The growth and activity of bacteria capable of degrading the chlorinated ethylenes can be stimulated by supplementing the water with aliphatic hydrocarbons and oxygen. However, delivering adequate quantities of these insoluble gases to the aquifer is a formidable engineering challenge.

*JTW, R.S. Kerr Environ. Research Lab., U.S. EPA, P.O. Box 1198 Ada, OK 74820; SF, Bioremediation Systems, 1106 Commonwealth Av., Boston, MA 02215; PVR, Dept. Civil Engineering, Stanford U., Stanford, CA 94305.

Basic Biochemistry and Physiology Studies (Bioremediation Systems)
    The chlorinated ethylenes that are most commonly released to the
subsurface environment are tetra- and trichloroethylene. Frequently,
contaminated ground waters also contain considerable quantities of the
1,2-dichloroethylenes and vinyl chloride, which probably result from
an anaerobic biotransformation of tetra- or trichloroethylene in the
subsurface (Barrio-Lage et al., 1986; Wilson et al., 1986). Figure 1
depicts the sequence. Chlorine atoms are replaced one at a time with
hydrogen atoms to produce first trichloroethylene, then cis- and trans-
dichloroethylene, then vinyl chloride. The dichloroethylenes or vinyl
chloride can accumulate, or they can be further transformed to products
that are nonvolatile or nonchlorinated (Vogel and McCarty, 1985).
1,1-Dichloroethylene is also produced by a nonbiological transformation
of 1,1,1-trichloroethane (T. Mill and W. Haag, SRI, Menlo Park, CA; T.
Vogel and P. McCarty, Stanford U., Stanford, CA, personal communication). The 1,1-dichloroethylene might accumulate or be transformed to
vinyl chloride in anaerobic subsurface environments (Barrio-Lage
et al., 1986).
    Organisms in soil exposed to a mixture of natural gas in air, or
in a mixed culture enriched on methane, can metabolize trichloroethylene to carbon dioxide and cell material (Wilson and Wilson, 1985;
Fogel et al., 1986). The mixed culture could also degrade cis- and
trans-dichloroethylene, 1,1-dichloroethylene, and vinyl chloride (Fogel
et al., 1986); however, tetrachloroethylene is not degraded. The most
plausible intermediate for trichloroethylene and vinyl chloride degradation is their epoxides. These two epoxides rearrange or hydrolyze
very rapidly in water which makes them difficult to detect (Figure 1).
The epoxides of cis- and trans-1,2-dichloroethylene are more stable and
have been confirmed by mass spectral analysis to accumulate in cultures
actively degrading the dichloroethylenes. The likely hydrolysis or
rearrangement products of the various epoxides, such as trichloroacet-
aldehyde, chloroacetaldehyde, and chloral were rapidly degraded in the
mixed culture (data not shown).

Goal of Stanford University's Field Demonstration
    In cooperation with the U.S. EPA and the U.S. Navy, a research
team within the Environmental Engineering and Science Group at Stanford
is conducting a field evaluation of trichloroethylene co-oxidation at a
site on Moffett Naval Air Station at Mountain View, CA (Figure 2). The
site is within the Santa Clara Valley, which is often called "Silicon
Valley" because of the local concentration of electronics and other
high-tech industries. Ground-water contamination with chlorinated
solvents is common in the region. The general approach is as follows:
1) choose a representative demonstration site based on available information about regional hydrology and geochemistry, and considering
institutional constraints; 2) survey the geology of the site through
analysis of cores, the hydrology through pump tests, and the geochemistry and extent of contamination through water analyses; 3) construct
a system of wells for injection, monitoring, and extraction of waters
moving through a shallow aquifer at the site; 4) devise an automated
system to sample the water moving through the aquifer and subject it to
chemical analysis; 5) observe the movement of a bromide past the wells
under the natural gradient as a tracer for the direction and velocity
of local ground-water flow; 6) compare the movement of trichloroethylene relative to the tracer to account for sorption and removal through

other pre-existing processes; 7) promote the growth of native methane-utilizing microorganisms by injecting dissolved methane and oxygen; and, 8) determine the removal of trichloroethylene resulting from the metabolism of the injected methane and oxygen.

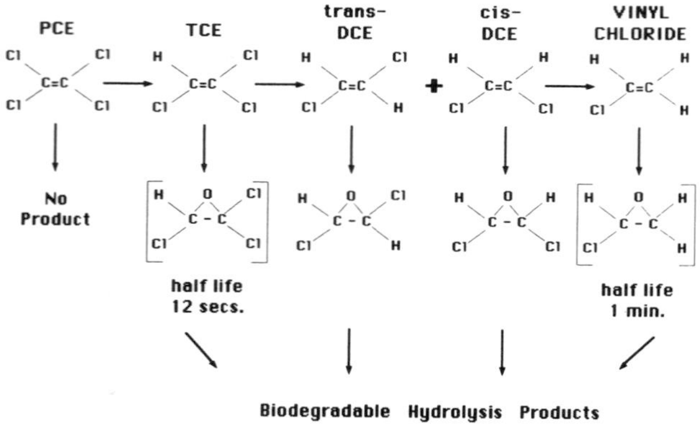

Figure 1. Biotransformation of chlorinated ethylenes and their breakdown products. PCE is tetrachloroethylene (also known as perchloroethylene), TCE is trichloroethylene, and DCE is dichloroethylene. The pathway moving to the right of the figure is a sequential reductive dechlorination that may occur in anaerobic environments. The pathway moving down the figure is a co-oxidation of the ethylenes that occurs during the aerobic metabolism of gaseous alkanes.

Design of Stanford University's Field Demonstration

The demonstration is being conducted in a shallow, semiconfined alluvial aquifer. The permeable zone consists of a layer of poorly sorted silty sand and gravel approximately 4 to 6 feet thick, bounded above and below by clay. An array of injection, monitoring, and extraction wells was installed to permit sampling at horizontal intervals of 3 to 6 feet (Figures 2 and 3). In order to achieve real-time control and interpretation of an experiment as it is in progress, the data-acquisition system samples and analyzes the water from individual wells continuously for several weeks or months. Typically, five points are sampled: the injected water, three monitoring wells, and the water in the extraction well. Each individual well is sampled every 2.5 hours. The waters are analyzed for oxygen using a probe with a detection limit of 0.1 mg/liter; for pH with an electrode; for bromide, chloride, nitrate and sulfate through ion chromotography with a detection limit of 0.5 mg/liter; for trichloroethylene, Freon 113 and 1,1,1-trichloroethane by gas chromotography using an electron capture detector at a detection limit of 1.0 µg/liter; and for methane using

gas chromatography and a flame-ionization detector at a detection limit
of 0.2 mg/liter. As far as possible, the samples are free of artifacts
due to losses or mixing in the sampling lines. The high density of
data allow accurate calculations of mass balance, which is required for
an unequivocal estimate of biodegradation in situ.

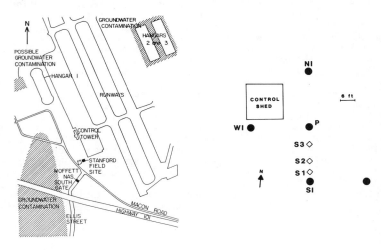

Figure 2. Location of Stanford University's field demonstration
of trichloroethylene biodegradation. A vicinity map is on the
left; the map of the demonstration site is on the right.

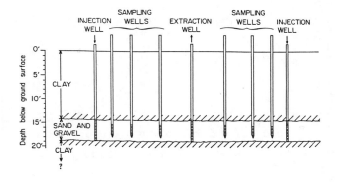

Figure 3. Section view of well layout.

Preliminary Field Results

The field facility is fully operational, and the aquifer has been characterized with respect to transport of water using a bromide tracer as well as transport of oxygen (Figures 4 and 5). During these experiments, water was injected at 1.0 liter/min in well SI and extracted at 8 liter/min from the central extraction well. The concentration of bromide and oxygen was monitored in wells S1, S2, and S3. The data shown represent a tightly spaced temporal record of the concentration response over ten days, with approximately ten samples per day. Data of this kind make it possible to infer the residence time distributions of fluid along the paths leading from the injection well to the various sampling wells under conditions representative of the demonstration experiments. The average residence times so estimated are on the order of 6 to 20 hours. The data also reveal that breakthrough at the monitoring wells is less than complete, a symptom of partial capture of injected solute by the regional ground-water flow, the velocity of which has been shown to be surprisingly high, on the order of several meters per day. Other preliminary experiments have shown that there is no significant disappearance of methane in the absence of dissolved oxygen and no significant disappearance of oxygen in the absence of methane. The transport data are being analyzed to determine the influences of advection, dispersion, and retardation. If the influence of these processes are accurately defined, then a mass balance can be used to calculate removals of trichloroethylene in subsequent experiments on biostimulation.

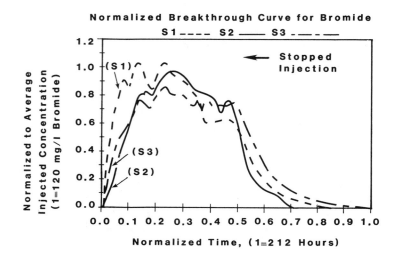

Figure 4. Breakthrough of bromine tracer in an injection test. S1, S2, and S3 are monitoring wells depicted in Figures 2 and 3.

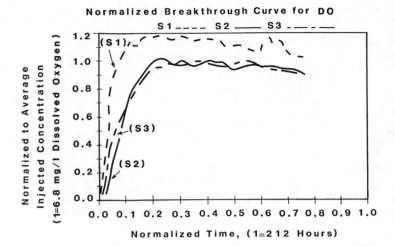

Figure 5. Breakthrough of oxygen in an injection
test. S1, S2, and S3 are monitoring wells depicted
in Figures 2 and 3.

Preliminary Microcosm Study at R.S. Kerr Laboratory
During construction of the Moffett Field Demonstration Site, core
material was acquired from four bore holes, three of which later became
injection wells NI, WI, and SI (Figures 2 and 6).

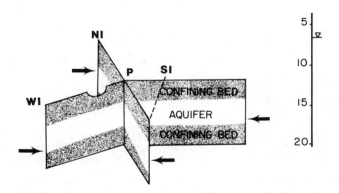

Figure 6. Original location of core material taken from
the field site for construction of microcosms. The vertical
scale is feet below land surface, north is to the top.

The core material was packed into glass columns (1.5 in. I.D. by 9 in. long); duplicate columns were packed from material from each bore hole. The columns were perfused with water from the Moffett Field Site that had been sparged with oxygen, and amended with an aqueous solution of both trichloroethylene and 1,1,1-trichloroethane, and an aqueous solution of either methane or propane. The pore volume of the columns was 90 ml. In each application cycle, 150 ml of water supplemented with oxygen, chlorinated hydrocarbons, and either methane or propane was applied to the column. Then the columns were sealed and incubated for ten days. Finally, the columns were perfused with 150 ml of deoxygenated Ada, OK tap water and 150 ml of effluent were collected for analysis of oxygen, methane or propane, trichloroethylene, and 1,1,1-trichloroethane.

Two sets of duplicate microcosms were perfused with water containing methane and two sets with water containing propane. The naturally occurring microbes in the core material acclimated and removed the methane or propane from the pore water by the end of the first application cycle. After two application cycles it became apparant that nonbiological processes were also removing the chlorinated organics, particularly trichloroethylene. To provide a control for this nonbiological removal, one column from each set of duplicates was poisoned by adding 0.1% sodium azide to the water used to perfuse the columns.

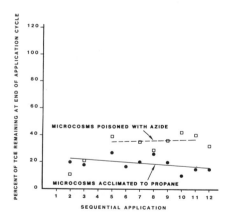

Figure 7. Disappearance of Trichloroethylene in aquifer material acclimated to degrade propane, compared to disappearance in material poisoned with sodium azide.

The disappearance of trichloroethylene in columns acclimated to propane is depicted in Figure 7. When the columns were poisoned with azide, roughly 40% of the material applied to the column was recovered at the end of the application cycle. Part of this material still remained on the core material after perfusion with the deoxygenated water that was used to force the sample from the column, and then was pushed from the column and wasted when the water containing oxygen, chlorinated organics, and hydrocarbon was introduced into the column at

the beginning of the next application cycle. However, a significant
fraction of the missing material probably sorbed to the solids. In the
columns acclimated to propane, roughly 20% of the material originally
applied to the columns was recovered, indicating that about one-half
of the remaining trichloroethylene was removed through biodegradation.
At the last cycle, the 90% confidence interval on the regression line
through the poisoned microcosms predicted that 30.8 to 41.0% of the
material applied in the last cycle was recovered from the microcosm.
The 90% confidence interval on the regression line through the living
microcosms predicted that only 12.3 to 18.7% of the trichloroethylene
still remained.

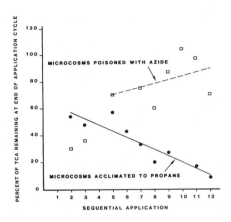

Figure 8. Disappearance of 1,1,1-Trichloroethane in aquifer
material acclimated to degrade propane, compared to its
disappearance in material poisoned with sodium azide.

The disappearance of 1,1,1-trichloroethane in the propane acclimated columns is depicted in Figure 8. Recoveries in the columns
poisoned with sodium azide were near 100%. The removals of 1,1,1-tri-
chloroethane in acclimated columns was poor in the first application
cycles, but by application twelve the removals averaged 90%. At the
last application, the 90% confidence interval on the regression line
through the poisoned microcosms predicted the 81.8 to 112% of the
1,1,1-trichloroethane applied to the microcosms still remained, while
the confidence interval on the line through the data from the living
microcosm predicted only 9.5 to 17% remaining. Over the entire experiment, the oxygen depletion in the microcosms acclimated to propane
were 54% and 64% of the theoretical oxygen demand of the propane
removed from perfusion water, while the oxygen depletion in the columns
that were acclimated to methane were 91% and 107% of the theoretical
demand. Perhaps biomass was accumulating in each application cycle in
the columns.

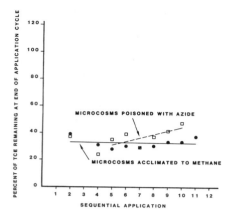

Figure 9. Disappearance of Trichloroethylene in aquifer material acclimated to degrade methane, compared to its disappearance in material poisoned with sodium azide.

Figure 9 depicts the removal of trichloroethylene in the columns that were acclimated to methane. As was the case before, the columns poisoned with azide showed extensive removal of trichloroethylene. At the tenth application cycle, the 90% confidence interval on the regression line through the data from the poisoned microcosms predicts that 20.8 to 34.9% of the trichloroethylene remained at the end of the cycle. By the end of the experiment, biodegradation was removing at most one fourth of the remaining trichloroethylene. The 90% confidence interval predicts that 19.5 to 29.6% of the material applied in the eleventh cycle still remained in the living microcosms. The methane acclimated columns received only one-third as much hydrocarbon as the propane acclimated columns, which may account for the less extensive biodegradation.

The recovery of 1,1,1-trichloroethane was near 100% in azide poisoned columns, while columns acclimated to methane removed roughly 40% of the 1,1,1-trichloroethane (Figure 10). The 90% confidence intervals predict that 64.1 to 98.6% of the trichloroethane remained in the poisoned microcosms after the tenth application cycle, while 55.8 to 75.6% of the material remained in the living microcosms after the eleventh application cycle. The extent of removal did not increase significantly with repeated applications, as was the case with columns acclimated to propane. At the end of the experiment, the two microcosms acclimated to propane, and one of their azide-poisoned controls, were repeatedly sampled in an attempt to wash trichloroethylene and 1,1,1-trichloroethane from the aquifer material. Each sampling required one hour. After four samples (equivalent to approximately six pore volumes), the columns were rested for four weeks to allow the concentrations of the chlorinated organics to come back to equilibrium. Then they were sampled another four times. It was difficult to wash out the sorbed trichloroethylene (Figure 11).

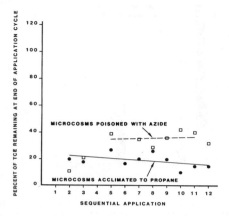

Figure 10. Disappearance of 1,1,1-Trichloroethane in aquifer
material acclimated to degrade methane, compared to its
disappearance in material poisoned with sodium azide.

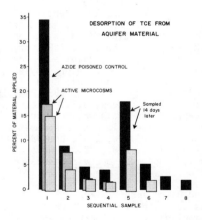

Figure 11. Slow desorption of Trichloroethylene from
aquifer material previously exposed to aqueous solutions
of Trichloroethylene.

Six pore-volumes of flushing reduced the concentration of trichloroethylene in solution to half its original value. Although
1,1,1-trichloroethane appeared to desorb more rapidly during flushing,
when the columns were rested the results were very similar to those for
trichloroethylene (Figure 12). More than a third of the original
concentration of 1,1,1-trichloroethane remained after six pore volumes
of flushing.

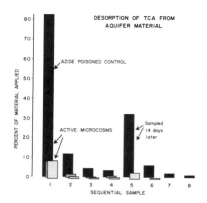

Figure 12. Slow desorption of 1,1,1-Trichloroethane from
aquifer material previously exposed to aqueous solutions
of 1,1,1-Trichloroethane.

References
1. Barrio-Lage,G.,Parsons, F.Z., Nassar, R.S., and Lorenzo, P.A.,
    "Sequential Dehalogenation of Chlorinated Ethenes," Environmental
    Science & Technology, Vol. 20, 1986, pp. 96-69.
2. Fogel, M.M., Taddeo, A.R., Fogel, S., "Biodegradation of Chlorinated
    Ethenes by a Methane-Utilizing Mixed Culture," Applied and
    Environmental Microbiology, Vol. 51, 1986, pp. 720-724.
3. Vogel, T.M., and McCarty, P.L., "Biotransformation of Tetrachloro-
    ethylene to Trichloroethylene, Dichloroethylene, Vinyl Chloride,
    and Carbon Dioxide under Methanogenic Conditions," Applied and
    Environmental Microbiology, Vol. 49, 1985, pp. 1080-1083.
4. Wilson, B.H., Smith, G.B., Rees, J.F., "Biotransformations of
    Selected Alkylbenzenes and Halogenated Aliphatic Hydrocarbons in
    Methanogenic Aquifer Material: A Microcosm Study," Environmental
    Science & Technology, Vol. 20, 1986, pp. 997-1001.
5. Wilson, J.T., and Wilson, B.H., "Biotransformation of Trichloro-
    ethylene in Soil," Applied and Environmental Microbiology,
    Vol. 29, 1985, pp. 242-243.

Acknowledgements and Disclaimer
     This effort has benefited from creative contributions from a large
number of people, including Lewis Semprini and Gary Hopkins at
Stanford; Doug Mackay now at UCLA; Jack Cochran, Michael Henson and
Marylynn Yates at Kerr Lab; and Margaret Fogel at Bioremediation
Systems. The work at Stanford was supported by the U.S. EPA through
CR-812220 with R.S. Kerr Lab. The work at Bioremediation Systems was
supported by Phase II of a Small Business Innovative Research Grant
from U.S. EPA Office of Exploratory Research, and by a National Science
Foundation Small Business Innovative Research Grant. Although the
research described in this abstract has been supported by the United
States Environmental Protection Agency, it has not been subjected to
Agency review and therefore does not necessarily reflect the views of
the Agency and no official endorsement should be inferred.

Physical and Chemical Treatment of Contaminated
Groundwater at Hazardous Waste Sites

By Thomas E. Higgins and Stephen Romanow[*]

INTRODUCTION

Groundwater contamination at hazardous waste sites is the driving force
behind most cleanup efforts. The significance of this environmental
threat has increased as the nation has become more dependent on its
groundwater resources. Most hazardous waste remediation efforts
involve some type of groundwater treatment.

In selecting an appropriate groundwater treatment system, three steps
are normally followed.

1. A groundwater investigation is performed to determine the types
    and concentrations of contaminants that are present and the waste
    matrices in which they occur.

2. Cleanup goals are established and translated into effluent
    standards.

3. Appropriate treatment technologies are selected, based on waste
    composition, cleanup goals, and knowledge of performance
    characteristics from prior experience with similar waste.

This paper elaborates on the process of how these steps have led to
the selection of physical and chemical treatment processes for
groundwater remediation at hazardous waste sites. Three case studies
are presented that illustrate actual applications of these
technologies. This information is condensed from feasibility studies
prepared for the U.S. EPA (CH2M HILL, 1984, 1986a, and 1986b).

TREATMENT SELECTION PROCESS

Groundwater Investigation

At a typical hazardous waste site, a groundwater investigation is
performed to determine types and concentrations of contaminants
present. The groundwater investigation includes installation of
groundwater monitoring wells and groundwater sampling and analyses.
Hazardous waste contaminants are classified as either inorganic or
organic, and the treatment methods that are proposed are very different
depending on which classification is found.

---

*Senior environmental engineer and environmental engineer,
respectively, CH2M HILL, P.O. Box 4400, Reston, Virginia 22090

Typical inorganic contaminants found in groundwater at hazardous waste
sites are antimony, arsenic, beryllium, cadmium, chromium, copper,
lead, mercury, nickel, selenium, silver, thallium, and zinc. Organic
contaminants primarily found at hazardous waste sites are classified as
volatile or semivolatile compounds. This distinction is made due to
their different physical and chemical properties and the different
treatment methods available for these two groups. Routinely, analyses
are run for 39 volatile and 68 semivolatile compounds.

Establishment of Treatment Goals

After groundwater contaminants are quantified, treatment goals are
established. Groundwater treatment involves removing contaminants from
the water and concentrating them in a reduced volume that can then be
safely disposed of. The objective is to produce the largest fraction
of treated water that can be discharged as a nonhazardous waste. Where
this treated water is discharged usually determines the required level
of treatment. There are typically four discharge options for treated
water:

1.   Discharge to a publicly owned treatment works (POTW)
2.   Discharge to a local surface-water body
3.   Recharge or reinjection into the groundwater
4.   Discharge to drinking water supply

Since each of these receiving waters typically has different quality
standards, this selection can have a significant impact on treatment
requirements. The EPA is currently developing a policy for discharge
of treated groundwater from hazardous waste sites to POTWs. Their
position is that discharge to POTWs will be allowed only if human
health and the environment are protected, if all relevant environmental
statutes are complied with, and if the receiving POTW has an acceptable
pretreatment program.

EPA has established pretreatment standards for a number of categories
of industries, i.e., categorical pretreatment standards. Since EPA has
not established categorical pretreatment standards for discharges from
hazardous waste sites, the most stringent categorical standards for the
contaminants present are typically applied. In most cases, the
Electroplating or Metal Finishing Categories (40 CRF Parts 413 and 433,
respectively) are applied, since they have effluent requirements for
both inorganic and organic contaminants. POTWs can apply local
pretreatment standards that can be more stringent than the federal
requirements. Effluent requirements for discharge to a POTW are
usually less stringent than for direct surface-water or groundwater
discharge, since additional treatment can be effected in the POTW.

Direct discharge of treated groundwater to a surface-water body near
the site would require a National Pollutant Discharge Elimination
System (NPDES) permit, with limits based on quality standards of the
receiving water body. Currently, no standards exist for discharge from
a hazardous waste site under the NPDES system, and each site is
evaluated on a "case-by-case" basis. Factors considered in issuing an
NPDES permit are the water use designation for the receiving water,

other wasteload allocations, relevant state and federal water quality
criteria, and other scientific data.

Aquifer recharge can be accomplished by land spreading or reinjection
of the treated groundwater. Water quality standards for underground
waters have been established by some states, usually based on drinking
water standards. Ordinarily, a state permit is needed to recharge
treated groundwater.

Direct discharge of treated groundwater to a potable water supply would
require treatment to meet drinking water quality standards. This type
of discharge is usually performed when a drinking water supply well is
contaminated. The treatment system is usually constructed adjacent to
the contaminated well.

TREATMENT PROCESS SELECTION

Inorganic Treatment Processes

Four treatment processes can be used to remove inorganic contaminants:
chemical precipitation, reverse osmosis, ion exchange, and electro-
dialysis. Of these, chemical precipitation is most commonly selected
for removal of inorganic contaminants from aqueous wastes at hazardous
waste sites, and it is discussed in detail.

Precipitation is a chemical treatment process in which dissolved
cations (positively charged soluble molecules) are combined with anions
(negatively by charged soluble molecules) to form insoluble salts.
Precipitation of dissolved metals (cations) is accomplished by increasing the concentration of the anion, of a slightly soluble metal-anion
salt, usually hydroxide (raising the pH).

Metal hydroxide precipitation is complicated by the fact that each
metal can have several soluble hydroxide species, resulting in an
increased solubility at a hydroxide concentration (pH)) greater than
optimal. The optimum pH for metallic hydroxide precipitation varies
for different metals as illustrated by Figure 1. The ability to meet
increasingly stringent effluent standards for metals has been limited
by the relatively high solubility of metal hydroxides and the lack of a
common pH of minimum solubility, as can be seen on Figure 1.

To improve removal further, sulfide precipitation can be used in place
of hydroxide precipitation. This fact is based on the theoretically
lower solubilities of some metal sulfides compared to those of metal
hydroxides. Figure 2 shows the theoretical concentrations of metals in
equilibrium with 10^{-6} molar total sulfide. Problems with metal sulfide
precipitation include the potential for producing toxic and noxious
hydrogen sulfide gas and the tendency to form colloidal precipitates
that are difficult to remove from solution.

Hexavalent chromium cannot be precipitated as simply as other metals
because it exists as chromate or dichromate anions. Both are negatively charged and cannot be precipitated as hydroxides unless
chemically reduced to the trivalent state. Chromium reduction,
followed by hydroxide precipitation with other metals, is the

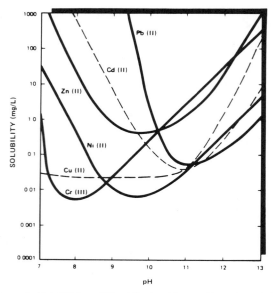

Figure 1. Solubilities of Metal Hydroxides as a Function of pH

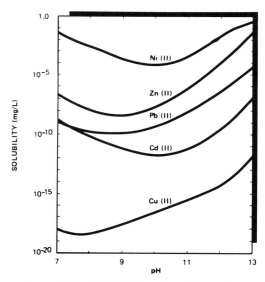

Figure 2. Metal Solubilities with 10^{-6} Molar Sulfides as a Function of pH

established method of treatment. Chromium reduction is normally conducted as a separate process step at a low pH (less than 3) using a reduced sulfur compound, such as sodium metabisulfite or sulfur dioxide, to contribute electrons to the reaction. Using sulfide as a reducing agent at a low pH is not recommended because it will form hydrogen sulfide.

Higgins and Marshall (1985) demonstrated that ferrous iron reduction of hexavalent chromium is sufficiently rapid at neutral to alkaline pH (7 to 10). The advantage of operating within this range is that the need for a separate acidic reaction is eliminated. However, additional sludge is produced due to the precipitation of iron hydroxide. Because of a coprecipitation effect, iron hydroxide precipitation can also improve removal of other metals.

Arsenic presents a removal problem similar to that posed by chromium, in that it is commonly present as an oxidized anion. Arsenic can be removed as part of coagulation. Lime treatment by itself is somewhat effective, although the most effective treatment has been addition of ferrous or ferric iron (at a ratio of 8:1) to the arsenic (at a pH of 5 to 6), followed by pH adjustment to 8 to 9. Iron to arsenic ratios greater than or less than 8:1 have been found to be less effective.

Reagents commonly used to effect the hydroxide precipitation include alkaline compounds such as lime or caustic soda (sodium hydroxide). Since lime is only slightly soluble, it is normally fed as a slurry, which makes it difficult to handle. Caustic soda is fed as a liquid because of its high solubility. Caustic soda is available as a 50 percent solution, although it is often fed at 25 percent because the more concentrated solution is subject to freezing. Caustic reacts more rapidly than lime.

The treatment chemicals may be added to a rapid mix tank, as shown on Figure 3. Because metal hydroxides tend to be colloidal in nature, flocculating and/or coagulating agents (polymers or metal salts) may also be added to aid in solids settling. These salts grow into particles of sufficient size that they can be removed from the water by physical unit processes, typically sedimentation and/or filtration followed by neutralization as shown on Figure 3. Table 1 summarizes typical design criteria for each of these unit processes.

ORGANIC TREATMENT PROCESSES

Removal of organic contaminants is typically accomplished using air stripping and/or carbon adsorption. Air stripping is used primarily to remove volatile organic compounds, while carbon adsorption is usually cost effective for removal of semivolatile contaminants.

Air Stripping

Air stripping is the process by which a compound that is dissolved in a liquid (usually water) is transferred to an air stream. The driving force for the transfer is related to a compound's vapor pressure in air relative to its solubility in water. Henry's Law Constant, a

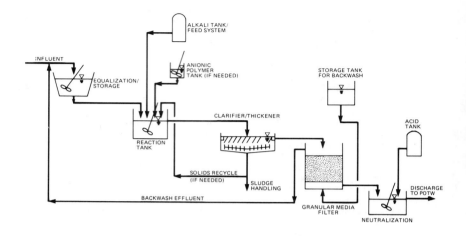

Figure 3. Typical Process Flow Diagram Using Hydroxide
          Precipitation Treatment

Table 1
INORGANIC TREATMENT DESIGN CRITERIA

	Chemical Precipitation (Rapid Mix Tank)	Sedimentation	Filtration	Neutralization
Detention Time	10-20 minutes	NA	NA	15 to 30 minutes
pH	9 to 10	NA	NA	7
Overflow Rate	NA	0.2 to 0.3 gpm/ft^2	NA	NA
Hydraulic Loading Rate	NA	NA	2 to 4 gpm/ft^2	NA

quantification of this ratio, is used to compute air stripping rates.
Removal efficiency is also highly dependent on temperature and air/water
ratios. Increases in Henry's Law Constant, temperature, or air/water
ratios produce increases in removal efficiencies, assuming that all
other factors remain constant.

Countercurrent packed towers are used for air stripping because of
their high mass transfer efficiency (see Figure 4). Water to be
treated is pumped to the top of the tower and cascades downward through

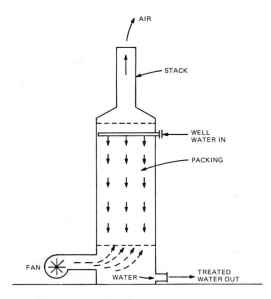

Figure 4. Packed Tower Schematic

packing media. The media retards the movement of the descending water
and provides a large surface area for contact between the water and the
air. Air is fan-forced in at the bottom of the tower and exits at the
top. Volatile compounds in solution are transferred from the water to
the air. When contaminated groundwater is being treated,
concentrations of compounds in the exiting air are usually well below
acceptable air emission limitations. If not, carbon adsorbers on the
exhaust air have been used.

There are numerous variables in the design of packed air stripping
towers. First, the flow rate of water to be treated, the volatile
compounds present, and the removal efficiencies of the compounds are
determined. The Henry's Law Constants for these compounds are then
looked up or estimated from handbook values for solubility in water and
vapor pressure. Design parameters include height and diameter of the
tower, air-to-water ratio, the type of packing to be used, and
acceptable pressure drop through the packing. Site limitations, such
as availability of land and height restrictions, can affect tower
diameter and packing depth.

Activated Carbon Adsorption

Whereas air stripping involves the removal (from water) of organic
compounds that are more attracted to air than to water, carbon
adsorption involves the attraction of nonvolatile organic compounds
that are more attracted to carbon than to water. "Activation" of

carbon is the selective burning of carbon particules to produce a
microporous structure of extremely large effective surface area on
which organic molecules can be adsorbed.

Factors that affect adsorption include the effective surface area of
the carbon, nature of the organic compound, pH, temperature, and
interference from mixed solutes. The primary design criteria for
carbon systems are surface loading rate and empty bed contact time.
Contaminated groundwater is normally treated at a surface loading rate
of between 2 and 7 gpm/ft^2 and an empty bed contact time of between 10
and 60 minutes.

Other important design criteria include bed volume and vessel
configuration. Bed volume is a function of contact time and design
flow rate. Sometimes bed volume is increased to reduce the frequency
of carbon regeneration. Carbon vessels can be configured for series or
parallel flow. Series flow is usually used for difficult-to-remove
pollutants where maximum contact time is desired. Parallel vessels are
used when large flow variations are expected or when large through-put
with a shorter contact time is desired.

Pilot testing is the most reliable method for acquiring design data,
although it is usually impractical for most situations. In addition,
for purposes of evaluating alternatives, information is needed prior to
pilot testing in order to verify that carbon is a viable option. This
makes operating data from previously operated or existing systems very
important for planning and design of carbon systems.

Comparison of Air Stripping and Activated Carbon Adsorption

The critical parameter affecting the selection of air stripping and
activated carbon technologies is the composition of the waste to be
treated. Air strippers operate on a percent removal basis, with
removal efficiency only minimally affected by influent concentration.
With increasing concentrations of influent waste, the effluent concentrations increase proportionally, and the total mass removed also
increases. As long as effluent concentrations remain acceptable,
treatment costs are unaffected. However, when effluent concentrations
are unacceptable, a larger stripping tower or increased air flow is
required to increase the removal efficiency. Activated carbon,
however, is mass loading limited, with carbon usage rate and operating
costs increasing directly with increased contaminant concentration,
although effluent concentrations may not increase.

The nature of the compound to be removed is very important as well.
Figure 5 presents a scatter plot showing relative ease of stripping
versus relative ease of adsorption for common toxic organics found in
groundwater. This diagram was derived from Henry's Law Constants, an
indication of stripping amenability, and the Freundlich K, an
indication of adsorptive amenability. These two parameters do not
relate directly, but plotting them together provides some insight into
the relative effectiveness of each process.

It is common to have groundwater contaminated with organic compounds
that are easy to air strip and compounds that are easy to be adsorbed

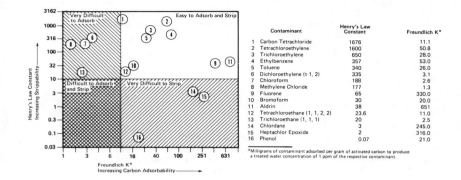

Contaminant		Henry's Law Constant	Freundlich K*
1	Carbon Tetrachloride	1676	11.1
2	Tetrachloroethylene	1600	50.8
3	Trichloroethylene	650	28.0
4	Ethylbenzene	357	53.0
5	Toluene	340	26.0
6	Dichloroethylene (t-1, 2)	335	3.1
7	Chloroform	188	2.6
8	Methylene Chloride	177	1.3
9	Fluorene	65	330.0
10	Bromoform	30	20.0
11	Aldrin	38	651
12	Tetrachloroethane (1, 1, 2, 2)	23.6	11.0
13	Trichloroethane (1, 1, 1)	20	2.5
14	Chlordane	3	245.0
15	Heptachlor Epoxide	2	316.0
16	Phenol	0.07	21.0

*Milligrams of contaminant adsorbed per gram of activated carbon to produce a treated water concentration of 1 ppm of the respective contaminant.

Figure 5. Strippability Versus Adsorbability

onto carbon. In these situations, air stripping and carbon adsorption are used in combination to take advantage of both. Figure 6 presents a typical process flow diagram for a combined organics removal system. Air stripping is provided first, since it is most efficient at removal of high concentrations and because it reduces the mass loading on the carbon, the limiting parameter on adsorption.

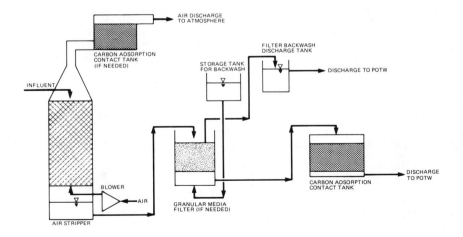

Figure 6. Typical Process Flow Diagram for Complete Removal

Air stripping can cause oxidation and precipitation of reduced metals (iron in particular). Filtration is typically used before the carbon system to prevent clogging of the carbon by oxidized metal hydroxides and other suspended solids that may be present.

CASE STUDIES

The Chisman Creek Hazardous Waste Site

The Chisman Creek site is located in the southeast portion of York County, Virginia, adjacent to Chisman Creek, a tributary of the Chesapeake Bay. The site consists of four fly ash disposal pits from 4.5 to 13.5 acres in area and up to 20 feet in depth. Contaminants found were principally inorganics, i.e., vanadium, nickel, and sulfates. The treatment process selected to treat this groundwater incorporates modifications to the typical process train to take into account the unusual waste composition.

The highest levels of contamination are found in the shallow groundwater within and immediately beneath the fly ash pits. Concentrations of total dissolved solids (TDS) and sulfate are approximately 10 and 100 times background levels, respectively. Vanadium and nickel are each about 1,000 times background. Arsenic, beryllium, chromium, copper, molybdenum, and selenium are at concentrations of up to 10 times background. Shallow groundwater downgradient from the fly ash pits also has levels of TDS and sulfate in excess of the Secondary Maximum Contaminant Levels (SMCL) of 500 and 250 ppm, respectively. Elevated concentrations of nickel, vanadium, beryllium, and molybdenum are also found in downgradient groundwater.

It is currently proposed to discharge treated groundwater directly to Chisman Creek. Therefore, the treated effluent would need to meet the Virginia surface-water quality criteria. Table 2 summarizes the contaminants of concern, maximum groundwater concentrations, and surface-water quality criteria.

Table 2
CHISMAN CREEK GROUNDWATER CONTAMINANTS

Contaminant	Maximum Influent Groundwater Concentrations (ug/l)	Standards* (ug/l)
Arsenic	60	50
Beryllium	60	None Available
Chromium	40	50
Copper	500	1,000
Molybdenum	440	None Available
Nickel	4,220	None Available
Selenium	30	10
Vanadium	21,600	None Available

*Standards adopted by the Virginia State Water Control Board

Metals precipitation, coupled with effluent filtration and solids handling capabilities, is the recommended treatment process for contaminated groundwater collected at the Chisman Creek site. Treated water would meet the requirements of an NPDES permit and, therefore, could be discharged to adjacent surface waters. The process flow diagram for the treatment scheme is shown in Figure 7.

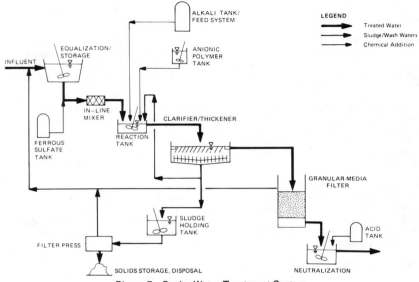

Figure 7. Onsite Water Treatment System, Chisman Creek Hazardous Waste Site

Relatively high flows are expected initially for gradient control or dewatering activities during construction of containment systems. These flows will be significantly reduced after the attainment of steady-state conditions. Estimated maximum and steady-state flows to the treatment facility are 125 and 50 gpm, respectively.

The equalization basin is to be designed with capacity for one-day storage of water when the flows to the plant decrease. The equalization tank will be mixed on a continuous basis to ensure that solids do not settle out.

Alkaline ferrous sulfate treatment was recommended for chemical reduction of hexavalent chromium and precipitation of vanadium. The resulting trivalent chromium will precipitate as a hydroxide along with other cationic metals. Vanadium will be precipitated directly as ferrous metavanadate. Ferrous iron will be added to the stream in excess of the stoichiometric amount.

Precipitation is carried out in a completely mixed reactor with a
detention time of 20 minutes. Lime is added to adjust the pH to 9.5.
Precipitation of ferrous metavanadate and other metal hydroxides occurs
optimally at about this pH. Ferric iron and other metals will precipitate as hydroxides. Aeration will be provided for oxidation of excess
ferrous iron, which precipitates as ferric hydroxide (assisting in
metals removal), and a polymer will be added to aid in coagulation.
Sludge from the clarifier is recirculated to the reactor to maintain a
solids concentration of 1 percent in the reactor to aid in vanadium
precipitation.

Removal of the precipitated metals will be accomplished in an inclined
plate clarifier/thickener. An effective overflow rate of 0.1 gpm/ft^2
was used for sizing the clarifier, based principally on the high solids
loading (approximately 1 percent). The sludge will either be recycled
to the reactor or pumped to a holding tank prior to dewatering.

Thickening will be required to achieve a reasonable solids recycle
rate. The thickener will be a "picket fence" type unit contained in
the sludge hopper of the gravity settler. Flow in the reaction tank
and clarifier will be continuous, regardless of the operation of the
rest of the plant, to prevent solidification of settled sulfate
sludges.

The clarified water will then flow to a granular media filter. A
continuously operating filter is recommended to eliminate the need for
backwash water storage, temporary storage of process water during
backwashing, and redundant units. The recommended filter loading rate
is 2 gpm/ft^2. Following filtration, the water will be neutralized with
sulfuric acid in a reaction tank of 15 minutes detention.

SEYMOUR RECYCLING CORPORATION (SRC) HAZARDOUS WASTE SITE

The Seymour site is located about 70 miles south of Indianapolis,
Indiana. This 14-acre site was operated as a waste chemicals process-
ing center from about 1970 until early 1980. Activities performed at
the site included reclamation, Chem-Fuel production, and destruction of
wastes by incineration. Over the years, hazardous wastes accumulated
on the site in 55-gallon drums, bulk tanks, and other containers.
About 50,000 drums, 98 bulk storage tanks, and many tank trucks that
were located on the site as of March 1980 have since been removed.
Hazardous substances and other wastes had leaked from these containers
onto the ground, causing soil and groundwater contamination.

A groundwater investigation showed that the shallow aquifer beneath the
site is highly contaminated with organic compounds, including
14 volatile and 6 semi-volatiles compounds. Compounds present in
greatest concentrations include trans-1,2-dichloroethene,
1,2-dichloroethane, 1,1-dichloroethane, and vinyl chloride. Shallow
groundwater offsite is primarily contaminated with volatile compounds.
Concentrations of 64,300 ppb of total volatile organic compounds have
been detected in the shallow groundwater up to 400 feet from the site.

It is proposed that treated groundwater be discharged to the Seymour
POTW. As mentioned earlier, there are currently no federal

pretreatment standards for discharge of treated groundwater from a
hazardous waste site to a POTW. The pretreatment standards for the
metal finishing and electroplating industries were used as the basis
for the proposed conceptual design. The metal finishing and
electroplating categorical standards include a limit of 2.13 ppm of
total toxic organics (TTO) for discharges greater than 10,000 gallons
per day. TTO is defined as the sum of all quantifiable organic
compounds that are listed in the standards and that are present in
concentration greater than 10 ppb. In addition to the federal
pretreatment standards, the City of Seymour's pretreatment regulations
limit phenols to 0.5 ppm.

In Table 3, selected contaminants found in the groundwater at the SRC
site are classified according to their amenability to being air
stripped or adsorbed. TTO and non-TTO contaminants are identified.

Table 3
SELECTED CONTAMINANTS IN GROUNDWATER AT SEYMOUR SITE

| | | Concentrations | |
| | | Average | Maximum |
Strippable Compounds	TTO*	(ppb)	(ppb)
Chloroethane	X	3,300	59,000
1,1-Dichloroethane	X	2,900	31,000
Methylene Chloride	X	1,900	32,000
Trans-1,2-Dichloroethene	X	15,100	240,000
1,1,1-Trichloroethane	X	2,800	93,000
Vinyl Chloride	X	1,900	40,000
Total Other TTO Compounds	X	2,800	78,300
Adsorbable Compounds			
N,N-Dimethyl Formamide		2,800	35,000
Phenols	X	100	4,800
Total Other TTO Compounds	X	100	1,500
Total Other Non-TTO Compounds		4,800	84,200
Low Strippability and Adsorbability Compounds			
Acetone		2,500	37,000
Tetrahydrofuran		3,500	52,000
Total Other Non-TTO Compounds		6,900	134,000
TOTAL		54,400	980,000
NON-TTO TOTAL		22,500	400,400
TTO TOTAL		31,900	579,600

*TTO denotes Total Toxic Organic compound, indicated by "X."

The process flow diagram for the treatment scheme proposed is the same
as that illustrated on Figure 6. Table 4 presents preliminary design
criteria used to size treatment units. The proposed treatment system
consists of air stripping followed by filtration and carbon adsorption.
The air stripper will be designed to achieve approximately 99.9 percent
removal efficiency. This system is intended to result in compliance
with pretreatment standards measured at maximum influent concentrations.
It will consist of a single air stripper, constructed so that the lower

Table 4
SEYMOUR PRELIMINARY GROUNDWATER TREATMENT DESIGN CRITERIA

Flow Rate = 150 gpm

	Air Stripper	Granular Media Gravity Filter	Carbon Adsorber
Packing Depth	45 ft	NA	NA
Air Flow Rate	2,000 cfm	NA	NA
Packing Pressure Drop	0.14 in/ft	NA	NA
Hydraulic Loading Rate	NA	2.5 gpm/ft^2	1.3 gpm/ft^2
Backwash Loading Rate	NA	15 gpm/ft^2	NA
Carbon Bed Depth	NA	NA	8 ft
Empty Bed Contact Time	NA	NA	30 minutes

portion of the tower will serve as a wet well. Stripped water is
pumped to the gravity filters.

Filtration is recommended prior to carbon adsorption, primarily because
of the iron concentrations (0.5 to 45 ppm) detected in the groundwater.
After contact with air, ferrous iron will be oxidized to ferric iron,
which precipitates as a hydroxide that has the potential for fouling
the carbon adsorbers. The filter will require periodic backwashing to
remove accumulated solids. It is proposed that the backwash be
discharged to the Seymour wastewater collection system along with the
treated groundwater. Filtered water will be discharged to a clear well
and pumped to the carbon adsorption system.

The carbon adsorption system will consist of two carbon adsorption
units in series in a lead-lag configuration. The system will include a
carbon transfer vessel into which spent carbon is transferred. In the
lead-lag configuration, the quality of effluent from the first (lead)

vessel is monitored for breakthrough of organics. When breakthrough
occurs, flow is switched so that the second (lag) vessel becomes the
lead vessel. After the exhausted carbon in the first vessel is
replaced, it is returned onstream as the lag vessel. This series
configuration ensures that a second contact stage is onstream to
protect against discharge of untreated or partially treated organics
when the first stage is exhausted.

"WEST COAST" HAZARDOUS WASTE SITE

The "West Coast" hazardous waste site is a generic name for a real
site on EPAs hazardous waste National Priority List. The 17-acre site
was operated as a bulk liquid disposal area from August 1956 to
November 1972. At least 34,000,000 gallons of organic and inorganic
industrial wastes from numerous generators was trucked to the site for
pond disposal. Site operations also included spray evaporation to
accelerate volume reduction.

Analytical data for the "West Coast" site show high concentrations of
metals in the groundwater at the original disposal area and immediately
downgradient (this groundwater is referred to as Stream A). Total
chromium, lead, and zinc levels are especially high. Groundwater
located at a different location, referred to as Stream B, shows much
lower metals concentrations. Table 5 summarizes the metal
concentrations detected in Streams A and B and their maximum effluent
concentrations. A similar pattern was observed in the concentrations

Table 5
GROUNDWATER CONTAMINANTS FOUND AT "WEST COAST" SITE
AND PRETREATMENT GOALS

Constituent	Stream A Groundwater Concentration (ppm) Average	Maximum	Stream B Groundwater Concentration (ppm) Average	Maximum	Maximum Effluent Concentration (ppm)
Arsenic	1	12	ND	ND	2
Cadmium	3	9	0.002	0.004	0.11
Chromium (T)	50	270	0.05	0.068	0.5
Copper	10	58	ND	ND	2
Lead	5	740	ND	ND	0.69
Mercury	0.01	0.063	0.004	0.3	0.03
Nickel	20	81	0.02	0.053	4
Zinc	50	130	0.05	0.13	2.61
Total Toxic Organics	15	32	2	3	0.58

of toxic organics in the groundwater. However, there are significant
concentrations of 1,2-dichlorobenzene, chloroform, and trichloroethylene
in the Stream B groundwater.

Discharge to a POTW is the selected method of disposal of treated
groundwater from the "West Coast" site. Effluent limitations
established by two regulatory bodies were considered: (1) a municipal
water district and (2) a county sanitation district. In addition, the
federal categorical pretreatment standards for the metal finishing
industry were also considered.

In order to assure acceptance of the treated groundwater, a
conservative approach was taken. It was proposed that for each
constituent, the most stringent requirement from the different sets of
limits discussed above be adopted.

A process flow diagram for the pretreatment system is shown on Figure 8.
Streams A and B are first treated separately and then combined for
organics removal. Stream A is expected to have an average flow of
20 gpm with a design peak flow of 50 gpm. Stream B is expected to
have an average flow of 40 gpm with a design flow of 80 gpm.
Preliminary design criteria used for sizing process equipment are
presented in Table 6.

The proposed process train for Stream A will provide removal of metals
and organics. The process train includes metal hydroxide precipitation,
sedimentation, mixed-media filtration, and carbon adsorption. Lime
was recommended for use in metals precipitation because lime sludges
demonstrate relatively good dewatering characteristics. A polymer
will be required to aid in flocculation.

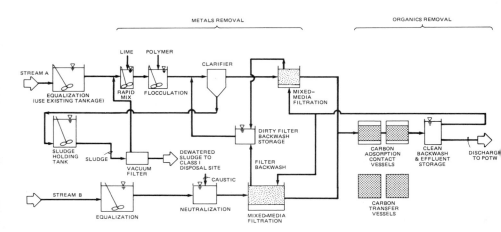

Figure 8. Process Flow Diagram Recommended Pretreatment
System, West Coast Site

Table 6
"WEST COAST" PRELIMINARY GROUNDWATER
TREATMENT DESIGN CRITERIA

Process	Parameter	Streams		
		A	B	A and B
Rapid Mix/ Neutralization	Detention Time (minutes)	1-5	30	NA
Flocculation	Detention Time (minutes)	20-30	NA	NA
Clarifier	Overflow Rate (gpm/ft^2)	0.2	NA	NA
Filter/ Adsorber	Hydraulic Loading (gpm/ft^2)	2.5-4	2.5-4	2.5
	Backwash Loading (gpm/ft^2)	15	15	NA
	Carbon Bed Depth (ft)	NA	NA	10
	Empty Bed Contact Time (minutes)	NA	NA	25

The clarified water will be passed through a mixed-media filter to
remove residual suspended solids that may otherwise plug the activated
carbon bed. The filter will require periodic backwashing to remove
accumulated solids. Dirty backwash will be returned to the clarifier.
Filtered Stream A will then be combined with Stream B and treated for
organics removal.

Treatment of Stream B will consist of equalization, neutralization,
mixed-media filtration, and carbon adsorption. A 4-hour equalization
period is recommended. The equalization tanks will be equipped with
mixers to keep solids in suspension. Flow equalization will be
followed by neutralization. Stream B has a pH of about 5. The pH
will be raised with sodium hydroxide so that after mixing with
Stream A, the pH of the combined streams will be 6.5 to 7.0. Sodium
hydroxide will be used instead of lime, because it is easier to handle
and does not have the solids content of lime.

Following mixed-media filtration, Stream B will be combined with
Stream A for activated carbon treatment. Two contact vessels in a
lead-lag configuration are to be provided as described in the SRC case
study. Treated effluent from the carbon contact vessels will be
hauled by tanker trucks to an industrial sewer for further treatment
at a POTW.

CONCLUSION

Contaminated groundwater treatment at hazardous waste sites is
generally a major component of the complete remedial cleanup action.

Groundwater treatment will continue to be an important factor in cleanup actions as the nation becomes more dependent on its groundwater resources. The groundwater treatment system selected is dependent upon effluent requirements and contaminants present.

The type of discharge will determine effluent goals and level of treatment. Four types of discharges are typically performed: (1) discharge to a POTW, (2) discharge to a local surface water body, (3) recharge or reinjection into the groundwater, and (4) direct discharge to a drinking water supply. Discharge to a POTW is desired because of the less stringent effluent requirements and additional treatment it offers.

The treatment processes selected will depend on the contaminants present and the effluent requirements. Inorganic contaminants are usually treated using a chemical precipitation process. Air stripping and carbon adsorption are usually the processes used to treat volatile and semivolatile organic compounds, respectively.

REFERENCES

CH2M HILL. Feasibility Study Report, Chisman Creek Superfund Site, York County, Virginia. EPA 83.3L37.0, August 1986.

CH2M HILL. Public Comment Feasibility Study Report, Seymour Recycling Corporation Hazardous Waste Site, Seymour, Indiana. EPA 50.5L01.0 and EPA 70.5L01.0, August 1986.

CH2M HILL. Fast-Track Remedial Investigation/Feasibility Study, "West Coast" Site, EPA 39.9M01 and EPA 39.9L010, May 1984.

Higgins, T.E., and B.R. Marshall. "Combined Treatment of Hexavalent Chromium with Other Heavy Metals at Alkaline pH." Proceedings of the Seventeenth Mid-Atlantic Industrial Waste Conference. June 23-25, 1985. Pp. 432-443.

The Impacts of CERCLA, RCRA and State Programs on Site Remediation: A Case Study

Kenneth Siet[1] and Kathryn L. Davies[2]

Abstract

Over the past decade the focus of environmental regulation both on the federal and state level has shifted from surface water to ground water. The major federal programs requiring ground water remedial response actions include RCRA and CERCLA. On the state level New Jersey law makers have enacted a number of statutes which regulate ground water. The various state and federal programs are evolving as distinct programs with separate goals. While all of these programs include requirements for ground water remedial actions, each has developed separate approaches to deal with the problem in response to each programs distinct statutory limits or goals.

The case study presented is of a complex facility with ground water contamination problems from numerous sources. The facility is on the NPL and is also subject to RCRA and HSWA corrective action requirements. Additionally the facility is subject to regulation under the more stringent state programs. The case study demonstrates how regulatory requirements drive technical decision making for ground water remedial actions. Decisions affected include such issues as the appropriate level of QA/QC all the way up to the major issue of "How Clean is Clean?". It also points out the dilema of having overlapping and conflicting regulatory programs at the same site.

With the increasing awareness of contamination problems in ground water and the fact that ground water provides over 50% of this country's potable water needs, it is not surprising that ground water protection and remediation has risen to a top priority for both federal and state lawmakers. The initial goal of recent regulatory programs has been to regulate the discharge of pollutants to ground water to prevent future pollution problems while concurrently requiring remediation of existing pollution problems. Most prevention programs are implemented through a permitting process whereby the facility owner/operator must institute minimum technology requirements for the operation and monitoring of units which may potentially or actually discharge to ground water. Prevention programs are designed to either

---

1. Section Chief, NJDEP, 401 E. State Street, Trenton, N.J.
2. USEPA Region III,841 Chesnut Street,Philadelphia, PA.

completely prevent a discharge or to limit the discharge to an environmentally safe level. Remediation programs are implemented through the permitting process or through state or federal enforcement actions which require responsible party ground water cleanups. If a responsible party does not exist or if the responsible party refuses to institute the required remedial action, public funds are available to ensure cleanup.

An industrial facility located in New Jersey may be subject to various federal and state ground water contaminat prevention and remediation programs. The two major federal programs which currently address ground water contamination programs are the Resource Conservation, Recovery Act of 1976 (RCRA) and the associated Hazardous and Solid Waste Amendments of 1984 (HSWA) and the Comprehensive Environmental Response, Compensation and Liability Act (CERCLA). RCRA is a "cradle to grave" waste management program. The statutory limitations of RCRA and HSWA are restricted to the prevention of discharges and remediation of environmental problems caused by, or associated with, strictly defined "solid waste" management activities. CERCLA is an environmental response program whose goal is to identify and remediate environmental problems caused by the mismanagement of hazardous substances. CERCLA actions are triggered in response to a verified release. CERCLA cleanup authorities are broader than those under RCRA whereby CERCLA may require remedial actions for environmental problems caused by either waste or product.

Pursuant to the statutory requirements of RCRA, regulations concerning ground water were promulgated on July 26, 1982. The regulations were specific to facilities with currently operating land disposal units (surface impoundments, land treatment areas and landfills) used to treat, store or dispose of specific waste types. In accordance with the rules, each of the RCRA regulated units at a facility was to have a ground water detection monitoring system capable of immediately detecting any discharge of pollutants from the unit. If a unit was found to be leaking, the regulations required that corrective actions be taken to remediate releases from that unit. Hence with the invention of the term "regulated unit", EPA's piecemeal approach to ground water pollution problems was initiated under the RCRA program. The scope of these initial rules was limited to ground water releases caused soley by a RCRA regulated unit. At the time of promulgation of this rule it was estimated that less than one percent of the land disposal units in New Jersey met the definition of a RCRA regulated unit.

Along with the requirement to institute corrective action under RCRA, EPA was obligated to address the issue of ground water cleanup standards. The original RCRA ground

water corrective action requirements (contained in 40CFR
Part 264 Subpart F) envisioned cleanup to background in
most cases. This concept came out of the assumption that
the Subpart F cleanup to background standards would apply
to primarily new land disposal units which met the strict
minimum technology standards of double liners and leak
detection systems. It also assumed that an adequate
detection ground water monitoring system was in place to
immediately detect any leakage at the limit of the waste
management area. It was hoped that the minimum technology
requirements, collectively with the detection monitoring
program, would guarantee that ground water problems would
be detected early and that the cleanup of ground water to
background values would be a realistic goal. This
approach may work at new land disposal units; however, it
is probably an unattainable goal at existing units or at
closed or abandoned solid waste management units. Neither
cost effectivness, or technological feasibility, are used
in setting cleanup standards. The cleanup standard is
based on the background quality of ground water as
measured directly upgradient of the unit.

The limitations of RCRA were recognized by Congress, and
the Hazardous and Solid Waste Amendments of 1984 (HSWA)
were enacted in attempt to correct some of the shortfalls
in the original RCRA program. In terms of ground water
contamination prevention and remediation, the HSWA
requirements greatly expanded the scope of the regulator's
authority. To aid in the prevention of ground water
contamination, HSWA set new minimum technology
requirements for land disposal units and required the
facilities to institute waste minimization programs. The
HSWA provision for the facility to address continuing
releases from solid waste management units, previously
excluded from regulation under the old RCRA, expanded the
regulator's authority to utilize a more "facility-wide"
approach for corrective actions. Even with the amendments
RCRA corrective action rules continue to focus on
releases solely from identifiable solid waste management
units.

Even with the expansion of the RCRA program by HSWA there
are major gaps in RCRA's approach to ground water
protection and remediation at operating industrial
facilities. The emphasis by RCRA on individual ground
water programs for each waste management unit at a
facility assumes that releases from individual units are
discernible. Additionally it assumes that all ground water
pollution emanates from waste management units and there
are no contributions of pollutants from "sloppy
housekeeping", spills, production areas, or other
uncontrolled events. Experience on the state level has
shown that this is not often the case. Individual plumes
at a facility typically mix over time to create a
composite plume from a variety of sources. In a small

highly industrialized state such as New Jersey,
neighboring facilities often create regional ground water
pollution problems.

The approach in preventing and remediating ground water
contamination under the RCRA program varies significantly
from the approach used to address the same issues in the
CERCLA program. While RCRA's goal is one of waste
management, CERCLA's main goal is protection of human
health and the environment from releases of hazardous
constituents from uncontrolled sources. The program is
based on a risk-based, cost effective cleanup strategy.

With the discovery of sites such as Love Canal in New York
and the realization that RCRA alone could not correct all
of the environmental problems associated with the
mismanagement of hazardous substances, Congress enacted
the Comprehensive Environmental Response, Compensation and
Liability Act (CERCLA). The original intent of CERCLA was
to implement remedial actions for sites which pose a
threat to human health and/or the environment as a result
of uncontrolled releases. CERCLA was initially envisioned
as a corrective action program designed to use public
funds to remediate abandoned sites or sites where the
owner/operator is not considered to be a responsible
party, able to properly administer and/or fund the
necessary cleanup.   CERCLA has the enforcement authority
to require responsible parties to implement remediation
actions through administrative or court imposed orders.

CERCLA is neither a regulatory program nor a preventative
program for ground water contamination. In addition to the
notification requirements in Sections 102 and 103 of the
Act, CERCLA has only limited requirements for the
treatment, storage or disposal of hazardous waste or
substances. Its remedial action requirements are based on
site specific risk assessments.   Its broad authority
allows site-wide remedial actions to include production
areas as well as waste disposal areas. CERCLA's ground
water remedial action requirements are based on a cost-
effective, risk assessment approach to corrective action.
Cleanup standards are determined on a site specific basis
based on the current exposure pathways to human
populations. The remedial options chosen must limit or
control exposure of the human population to the
contaminated ground water. This approach allows for a
high degree of flexibility in choosing the appropriate
remedial action alternative and in determining cleanup
standards for ground water. This differs greatly from the
stringent regulatory approach required under RCRA. Under
the CERCLA program, emplacement of an alternate water
supply to the affected population may constitute a viable
remedial action in lieu of a ground water cleanup.

An industrial facility was chosen to illustrate how

regulatory programs can interact, overlap and ultimately
differentially drive various clean up mechanisms. The
facility is complex both from the technical and regulatory
point of view. Groundwater investigations beginning with
geophysical surveys and expanding to include over 200
monitoring wells have documented the presence of extensive
ground water contamination problems resulting from
facility operations. Individual onsite wells have reported
values of over 25,000 ppb of total priority pollutant
volatile organics.

The facility occupies approximately 1,300 acres of land in
the coastal plain of south central New Jersey. Slightly
over 300 acres of the track is developed with
approximately 30 buildings comprising the manufacturing
operation. The facility is located in an area with both
residental and commercial development. There are
approximately 20 municipal water supply wells located to
the northeast of the facility. The site is relatively
flat sloping slightly in an easterly direction with the
major topographic feature being a river located to the
east of the facility and bordering the facility to the
northeast. There is a sharp dropoff to the river at this
northeast boundary. The river is fed primarily from
ground water discharge.

The facility began operation over thirty years ago. The
facility is engaged in the manufacturing of organic
chemicals. A permitted wastewater treatment plant with a
capacity of over 7 million gallons per day treats the
facility's process waste streams. The facility has a
number of active as well as inactive waste management
units on site. These include: an active landfill, an old
landfill (drum disposal area and lime sludge disposal
area), a filter cake disposal area, several backfilled
lagoons and two active equalization basins.

Hydrogeologically, the facility is situated within the New
Jersey Coastal Plain consisting of over 2000 feet of
unconsolidated sediments. The two major stratigraphic
units of concern are the Cohansey sands and the underlying
Kirkwood formation. The upper Kirkwood formation serves
as a aquitard and limits the vertical flow of water from
the Cohansey. There have been two confined water bearing
zones identified within the lower Kirkwood formation in
the vicinity of the facility. Ground water monitoring has
consistently shown that both zones are clean. The water
table aquifer is in the Cohansey sands and is used as a
significant potable water supply within the region.
Groundwater flow within the Cohansey is generally west to
east towards the adjacent river. Current piezometric
studies indicate that the river serves as a downgradient
hydrogeologic barrier for the Cohansey aquifer units. A
series of investigations at the facility have revealed a
contaminant plume within the Cohansey sands. The plume is

approximately 300-400 acres in extent and consist of
primarily volatile organic chemicals (see figure 1).

The characterization of the plume was a lengthy process
driven by federal and state regulatory programs. The
initial step in the process was the facility's
registration of the old landfill with the state. The old
landfill clearly had the potential to discharge pollutants
to the ground water and under a State Administrative
Consent Order (ACO) issued pursuant to the state's Water
Pollution Control Act, the facility was required to
install monitoring wells downgradient of the landfill.
The wells detected significant concentrations of volatile
contaminants in the ground water and under the auspices of
the ACO, the facility was directed to instigate a pumping
program to prevent the further movement of contamination
off-site. Several low rate pumping wells were installed
downgradient of the landfill. The facility was
additionally required to implement source control
measures. Additional wells were required around the two
equalization basins to comply with the interim status RCRA
land disposal regulations. In accordance with the RCRA
rules the facility installed a detection ground water
monitoring system around the basins.

Based upon the existing ground water data and the
proximity of the municipal well fields, the site was
ranked, placed on the NPL and subject to CERCLA corrective
action requirements. At the time of the ranking, only
units and production areas which were not regulated by
RCRA or by the State ACO were considered. EPA decided
that a Remedial Investigation/Feasibility Study (RI/FS)
study was necessary to determine the nature, extent and
sources of the ground water contamination. EPA also
decided that the facility was not a responsible party who
would be able to properly conduct the RI/FS and EPA
contracted a private consulting firm using monies provided
by the facility. As shown on Figure 1. Areas 3,4,6,7 and 8
were subject to investigations and corrective action under
the CERCLA program. Areas 1 and 2 remained subject to
regulation solely under the RCRA rules. While Area 5 was
regulated exclusively by the provisions of the state ACO.

Concurrently with the EPA sponsored RI/FS, the facility
decided to conduct its own RI for evaluating all areas of
potential contamination, including specific waste and
production areas. The facility's RI was broader in scope
than either the federal or state RIs since it was not
being implemented in accordance with, nor subject to the
regulatory limitations of any one regulatory program.

Soon after both RIs were initiated, the state determined,
based on ground water data generated by the RCRA detection
monitoring program, that the equalization basins were
leaking and impacting ground water quality. In accordance

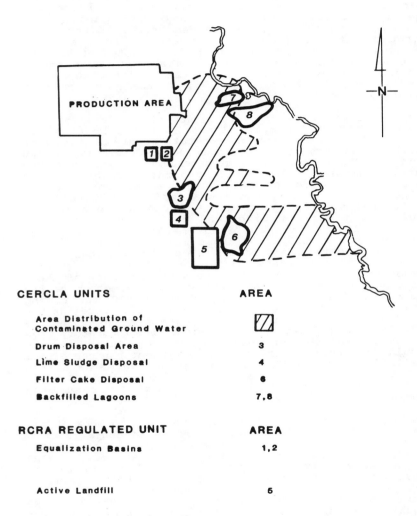

CERCLA UNITS   AREA

Area Distribution of
Contaminated Ground Water

Drum Disposal Area   3

Lime Sludge Disposal   4

Filter Cake Disposal   6

Backfilled Lagoons   7,8

RCRA REGULATED UNIT   AREA

Equalization Basins   1,2

Active Landfill   5

FIGURE 1   Simplified Map of Facility

with the RCRA requirements, the facility had to conduct a
limited study to determine the nature and extent of
contamination emanating from the RCRA regulated units.
This ground water quality assessment program showed that
the plume of contamination moved under and slightly past
the river and was then discharged back into the river
through natural ground water flow. The facility
implemented a pumping program which consisted of two low
rate pumping wells located immediately down gradient of
the units. The pumping program was designed to have a
limited zone of capture to include the down gradient
boundry of the units.

Within two years the facility had completed its site wide
RI and EPA had completed the RI for all units not
regulated through the RCRA and state program. Problems
existed with each of the different RIs. Conclusions drawn
from the EPA sponsored RI were dependent on limited ground
water monitoring information generated by one or two
sampling events from 60 CERCLA approved monitoring wells.
Although the facility had installed approximately 150
additional monitoring wells the CERCLA EPA sponsored RI
could not utilize this ground water monitoring data
because; 1) the data could not meet the strict QA/QC
requirements of CERCLA, 2)these wells did not meet the
well construction standards of EPA's CERCLA guidance, and
3) the facility, determined by EPA to be a "non-
responsible party", had taken the samples. Many of these
additional wells were installed at the direction of the
State and EPA to meet RCRA and state ACO monitoring
requirements. Even though CERCLA would not approve the use
of this data, the ground water samples taken by the
facility and analyzed by a state certified laboratory show
consistent values over a period of three years and met the
regulatory requirements of RCRA. Without the use of all
available wells, all historical ground water data, and
without consideration of all units on site, the CERCLA RI
was limited both in its scope and the resulting
conclusions.

The RI conducted by the facility attempted to look at all
available data and all potential sources of contamination.
However, not all of the sampling results from the CERCLA
RI were accessable to the facility. Therefore, the RI
conducted by the facility was also limited in its
conclusions. The separate study conducted by the facility
for the RCRA regulated units was the most restrictive in
nature because it was limited strictly to the two RCRA
regulated units. Based on a statistical comparison between
upgradient and downgradient wells, the basins were assumed
to be the sole sources for the ground water contamination
in the immediate area. This conclusion was appropriate
given the narrow scope of RCRA required study. However,
under the more comprehensive RI conducted by the facility,
the adjacent production area as well as waste management

units 7 and 8 were shown to be likely contributors to the plume.

The choice of the appropriate corrective action and the cleanup criteria for the contaminated ground water will be driven by the requirements of the regulatory program. CERCLA corrective actions are based on cost effectiveness and analyses of risks to human health; the actions are not designed to clean up the aquifer to protect ground water. At this site the river serves as an effective hydrogeologic barrier to the down gradient movement of contaminated ground water. There is limited impact from the contaminants to the river because of their volatility. Although the plume of contamination has migrated into a residential area, there are no down gradient potable wells between the facility and the adjacent river. With this scenario a viable CERCLA corrective action would be to confine the sources with a cap and/or remove the sources and monitor ground water. Ground water remediation may not be needed since the river acts as a natural barrier all and existing residences lying over the plume are served by municipal water lines.

The RCRA investigation concluded that there was ground water contamination caused by the RCRA regulated units. Pursuant to the RCRA regulations the facility would be expected to implement corrective actions to clean up that ground water which was impacted by the RCRA units to established "concentration limits", which in most instances are background values. The regulations specify that pollutants in the ground water downgradient of the unit are assumed to be emanating from the unit unless it can be proven otherwise. The facilities RI concluded that the production area is contributing to the "RCRA" plume. If the facility could distinguish between pollutants from the production area and from the regulated unit, the facility would only be required to clean up the pollution from the regulated unit to meet the requirements of the RCRA rules. If the sources of pollution could not be individually identified, a clean up program would be mandated to continue until background values were reached. Without a remediation program in the production area, the clean up program could continue indefinitely.

The site wide RI conducted by the facility identified a large composite plume of volatile organic contamination emanating from numerous sources including both RCRA and CERCLA units. The facilities RI suggests that complete source removal may be impossible. Some of the organic pollutants found are dense non-aqueous phase liquids (DNAPLs) which tend to move vertically through ground water as globules and pool in various depressions within the aquifer. There is no known method for detecting the location of these pools of DNAPLs within the aquifer. Therefore, these DNAPLs may provide a continuing source of contamination unless pumping wells fortuitously penetrate every pool of these chemicals.

At this site neither federal regulatory program provided for the evaluation of all sources of contamination, nor all plumes of contamination emanating from the facility. Ground water investigations and response actions were designed based upon the goals and limitations of the individual programs. Under the current RCRA rules a land disposal unit may be subject to strict minimum technology liner and leak detection system requirements while an adjacent unit holding similar pollutants may have no minimum technology requirements. A facility operating a cleanup under CERCLA requirements may not be required to remediate ground water while an adjacent facility subject to RCRA and having impacted the ground water to a similar or even lesser extent may be required to cleanup ground water to natural background levels.

Ground water is a resource that requires sound management and planning. Rational decisions for it's protection and remediation must be based on a coherent resource oriented footing. The federal government has taken this approach in it's regulation of surface waters under the NPDES program. New Jersey is attempting to incorporate it's ground water program within it's overall water resource program. This resource oriented approach allows the state to regulate discharges to ground water regardless of the source. It allows facility wide instead of unit by unit decision making. It also allows for regional aquifer management and planning which takes into account both present and future ground water use. Recent reauthorizations of the RCRA and CERCLA programs greatly expanded the scope of both program's abilities to address ground water remediation. However, the ground water aspects of these programs must be integrated with a sound ground water management and planning program.

CHAPTER 6.0

LIST OF AUTHORS

Session No. 2-S2.1: Groundwater Contamination

J. David Dean
Woodward-Clyde Consultants
100 Pringle Avenue (Suite 300)
Walnut Creek, CA 94596

Dr. Norbert Dee (WH-550G)
Office of Ground Water Protection
US Environmental Protection Agency
401 M Street, S.W.
Washington, D.C. 20460

David R. Gaboury
Woodward-Clyde Consultants
100 Pringle Avenue (Suite 300)
Walnut Creek, CA 94596

John Gaston
Assistant to Commissioner for Hazardous Waste Management
Department of Environmental Protection, CN028
Trenton, NJ 08625

Marian Mlay
Office of Ground Water Protection
US Environmental Protection Agency
401 M Street, S.W.
Washington, D.C. 20460

Steven P. Roy
Ground Water Protection Department
DEQE, Division of Water Supply
Boston, MA 02108

Dr. Mahfouz H. Zaki
Suffolk County Department of Health Services
225 Rabro Drive East
Hauppauge, NY 11788

207

Session No. 9–S2.2: Monitoring and Detection

Kwasi Boateng
Roy F. Weston, Inc.
1 Weston Way
West Chester, PA

Dr. Olin Braids
Geraghty & Miller Inc.
125 E. Bethpage Road
Plainview, NY 11803

Edward Heyse
Air Force Engineering Services Center
Tyndall Air Force Base, Florida 32403
Dr. Edward Kaplan
Brookhaven National Laboratory
Upton, NY 11973-5000

Dr. Joseph Keely
Water Research Laboratory
Dept of Chemical, Biological and Environmental Sciences
Oregon Graduate Center
Beaverton, OR 97006-1999

Aldo T. Mazzella
US EPA/EMSL
Box 15027
Las Vegas, NV 89114-5027

Anne F. Meinhold
Brookhaven National Laboratory
Upton, NY 11973-5000

Ann M. Pitchford
US EPA/EMSL
Box 15027
Las Vegas, NV 89114-5027

Richard Schowengerdt
Geotechnical Project Manager
International Technology Inc
Milwaukee, WI 53224

Dr. Gregory Shkuda
Geraghty & Miller Inc
125 E. Bethpage Road
Plainview, NY 11803

Gisella M. Spreizer
Geraghty & Miller Inc
125 E. Bethpage Road
Plainview, NY 11803

Session No. 16-S2.3:   Control of Contamination

Dr. John Armstrong
Traverse Group, Inc.
2480 Gale Road
Ann Arbor, MI 48105

Robert M. Cohen
GeoTrans, Inc.
250 Exchange Place
Suite A
Herndon, VA 22070

Charles R. Faust
GeoTrans, Inc.
250 Exchange Place
Suite A
Herndon, VA 22070

Alexander J. Fazzini
OHM, Inc
P.O.B. 41
Windsor, NJ 08561

Ron Hoffer
Office of Ground Water Protection
US EPA
401 M Steet, SW
Washington, DC 20460

Dr. Martin Jaffe
University of Illinois at Chicago
School of Urban Planning & Policy (M/C 348)
P. O. Box 4348
Chicago, IL 60680

Dr. James W. Mercer
GeoTrans, Inc.
250 Exchange Place
Suite A
Herndon, VA 22070

Art Rosenbaum
OHM, Inc
P.O.B. 41
Windsor, NJ 08561

Anthony D. Truschel
GeoTrans, Inc.
250 Exchange Place
Suite A
Herndon, VA 22070

Session No. 23–S2.4: Renovating Contaminated Groundwater

Dr. K. C. Bishop III
Chevron, Inc.
575 Market Street
San Franscisco, CA 94105

Kathryn L. Davies
U. S. Environmental Protection Agency
Region III
841 Chestnut Street
Philadelphia, PA 19107

Dr. Sam Fogel
Bioremediation Systems
1106 Commonwealth Avenue
Boston, MA 02215

Dr. Thomas E. Higgins
Project Engineer
CH2M–Hill
625 Herndon Parkway
Herndon, VA 22070

Dr. Paul V. Roberts
Dept. of Civil Engineering
Stanford University
Stanford, CA 94305

Ken Seit
NJ Bureau of Ground Water Quality Management
1474 Prospect Street
Trenton, NJ 08625

Dr. John Wilson
US EPA Robert S. Kerr Environmental Research Laboratory
P.O.B. 1198
Ada, OK 74820

SUBJECT INDEX

Page number refers to first page of paper.

AUTHOR INDEX

Page number refers to first page of paper.